T0282528

THE FOUNDERS OF SEISMOLOGY

THE FOUNDERS OF SEISMOLOGY

BY

CHARLES DAVISON, Sc.D., F.G.S.

Author of *A History of British Earthquakes*,
A Manual of Seismology

CAMBRIDGE
AT THE UNIVERSITY PRESS
1927

CAMBRIDGE
UNIVERSITY PRESS

University Printing House, Cambridge CB2 8BS, United Kingdom

Published in the United States of America by Cambridge University Press, New York

Cambridge University Press is part of the University of Cambridge.

It furthers the University's mission by disseminating knowledge in the pursuit of education, learning and research at the highest international levels of excellence.

www.cambridge.org
Information on this title: www.cambridge.org/9781107691490

© Cambridge University Press 1927

This publication is in copyright. Subject to statutory exception and to the provisions of relevant collective licensing agreements, no reproduction of any part may take place without the written permission of Cambridge University Press.

First published 1927
First paperback edition 2014

A catalogue record for this publication is available from the British Library

ISBN 978-1-107-69149-0 Paperback

Cambridge University Press has no responsibility for the persistence or accuracy of URLs for external or third-party internet websites referred to in this publication, and does not guarantee that any content on such websites is, or will remain, accurate or appropriate.

In Memory of

THE FOUNDERS OF SEISMOLOGY

and especially of those who have honoured me
with their friendship

JOHN MILNE

FUSAKICHI OMORI

and

ERNST VON REBEUR-PASCHWITZ

PREFACE

The present volume has its origin in three articles published in the *Geological Magazine* in 1921. These articles form the foundation of the second, fifth and tenth chapters, and I am indebted to the courtesy of the Editor, Dr R. H. Rastall, F.G.S., for permission so to make use of them. The writing of the remaining chapters is due to a suggestion made to me by the late Sir Archibald Geikie, O.M., F.R.S., and my one regret in connexion with this book is that he did not live to see the result of his kindly interest.

CHARLES DAVISON

Cambridge
November 1926

CONTENTS

ILLUSTRATIONS

[illegible]

ILLUSTRATIONS

NOTES

Portraits. Portraits of some of the Founders of Seismology will be found in the following books and journals:

Bertelli, T. *Riv. Geog. Ital.*, vol. 12, 1895, p. 5 of memoir.

Dutton, C. E. *Amer. Seis. Soc. Bull.*, vol. 1, 1911, opp. p. 137.

Humboldt, A. von. K. Bruhn's *Life of Alexander von Humboldt*, trans. by J. and C. Lassell, 1873, vol. 1, frontispiece; vol. 2, frontispiece and opp. p. 337.

Johnston-Lavis, H. J. *Bibliography of the Volcanoes of Southern Italy*, 1918, frontispiece.

Kikuchi, D. *Who's Who in Japan*, 1912, p. 362.

Lyell, C. *Life of Sir C. Lyell*, 2 vols., 1881, frontispiece to each volume.

Mercalli, G. *Ital. Soc. Sism. Boll.*, vol. 17, 1913, opp. p. 245.

Milne, J. *Amer. Seis. Soc. Bull.*, vol. 2, 1912, opp. p. 6; *Geol. Mag.*, vol. 9, 1912, opp. p. 337; *Ital. Soc. Sism. Boll.*, vol. 17, 1913, opp. p. 102.

Milne Home, D. *Edin. Geol. Soc. Trans.*, vol. 6, 1893, opp. p. 119.

Montessus de Ballore, F. de. *Amer. Seis. Soc. Bull.*, vol. 2, 1912, opp. p. 217.

Omori, F. *Amer. Seis. Soc. Bull.*, vol. 2, 1912, opp. p. 153.

Perrey, A. Dijon, *Ac. Sci. Mém.*, 1924, frontispiece to memoir.

Rossi, M. S. de. *Ital. Soc. Sism. Boll.*, vol. 15, 1911, opp. p. 10.

Suess, E. *Nature*, vol. 72, 1905, frontispiece.

Tacchini, P. *Ital. Soc. Sism. Boll.*, vol. 10, 1904, opp. p. 168.

References. In the footnote references, the abbreviated titles of journals are those adopted in the latest volume of the *Royal Society Catalogue of Scientific Papers*, or are formed on the same lines.

CHAPTER I

BEFORE MICHELL

1. I have supposed the birth of seismology to date from the middle of the eighteenth century, from the time when those who studied earthquakes drew their illustrations from contemporary records and no longer from the writings of Aristotle, Seneca or Pliny. From this point of view, the first Founder of Seismology would be John Michell (*c.* 1724–93), at one time Woodwardian professor of geology at Cambridge. Much of the material used in his memoir on earthquakes (1760) was, however, derived from two volumes, both published in 1757—*The History and Philosophy of Earthquakes* and Bertrand's *Mémoires Historiques et Physiques sur les Tremblemens de Terre.* I have therefore chosen these books as my starting-point, and they form the subject of the present chapter.

The other limit of my survey is less definite. It seemed clear that living writers should be excluded, and this rule, with one or two exceptions, has been followed. On the whole, I have confined myself to the history of seismology before its latest and most interesting development, so that, roughly, the period adopted may be said to end with the nineteenth century. The three chapters on Montessus (1851–1923), Milne (1850–1913) and Omori (1868–1923), however, naturally carry on the history for some years further; and, in the eighth chapter, a slight account is given of the foundation of the Seismological Society of America and of some recent work in the United States.

2. Two events, or series of events, led up to, if they did not suggest, Michell's great memoir, which forms the subject of the next chapter. The first was the remarkable group of earthquakes in England during the year 1750, the second the destructive Lisbon earthquake of 1755. In the "year of earthquakes," as the former year was called at the time, there were five strong shocks in this country, on 19 Feb. and 19 Mar. (N.S.) in London and the home counties, on 29 Mar. in Portsmouth and the Isle of Wight, on 13 Apr. in the north-west of England and the north-east of Wales, and on 11 Oct. in Northamptonshire and the surrounding counties. So great was the interest excited by them that, before the end of the year, nearly fifty articles were communicated to the Royal Society, in which the shocks were described or the causes and philosophy of earthquakes considered*.

To these shocks we are indebted for some of our earliest catalogues of earthquakes. Lists of British earthquakes appeared anonymously in the *London Magazine* and the *Gentleman's Magazine* for March 1750, the former recording 48 earthquakes and the latter 24. Later in the year, a third and more important catalogue was issued as a pamphlet—"*A Chronological and Historical Account of the most memorable Earthquakes that have happened in the World, from the beginning of the Christian Period to the present year* 1750; with an Appendix, containing a distinct series of those that have been felt in England, and a Preface, seriously addressed to all Christians of every Denomination." Though it was published anonymously as "By a Gentleman of the University of Cambridge," its author is known to have been the Rev. Zachary Grey (1688–1766), an antiquary of wide reading†. The main part of the pamphlet (pp. 8–44, 75–78) contains accounts of 61 destructive earthquakes, 14 of which occurred in China, Japan, Peru, etc. In the Appendix (pp. 45–74) are described 41 earthquakes felt in England from 974 to 1750.

Five years later, on 1 Nov. 1755, the city of Lisbon was destroyed by one of the greatest of recorded earthquakes. The curious seiches observed in our lakes and pools, the sea-waves that swept our southern coasts, and reported observations of the shock itself in various parts of England formed the subjects of many letters communicated to the Royal Society during November and the following months, 25 of them referring to observations in this country and 20 to those made in foreign lands‡.

* *Phil. Trans.* 1750, pp. 601–750.

† *Dict. Nat. Biog.* vol. 23, 1890, pp. 218–219. *A farther Account of memorable Earthquakes to the present Year* 1756 (Cambridge, 38 pp.), by the same author, describes some earthquakes, most of which had escaped his notice in 1750, as well as some letters on the Lisbon earthquake of 1755.

‡ *Phil. Trans.* vol. 49, 1756, pp. 351–444. Four other letters printed with the above refer to subsequent earthquakes.

JOHN BEVIS and *THE HISTORY AND PHILOSOPHY OF EARTHQUAKES*

3. The interest aroused by the Lisbon earthquake is responsible for the publication in 1757 of a remarkable volume on *The History and Philosophy of Earthquakes** by "A Member of the Royal Academy of Berlin." This is a collection of ten memoirs, some of them abridged, containing "the sentiments of the best naturalists as to their causes," in which, as the editor says, "he has retained entirely the facts, arguments and conclusions of the authors...without ever presuming to criticise any hypothesis, much less to obtrude one of his own." At the present time, the chapter of greatest value is one in which he has collected accounts of the Lisbon earthquake from the *Philosophical Transactions* and other "literary and authentic vouchers," arranging them under the places in alphabetical order.

After the lapse of more than a century and a half, it would have been difficult to discover the identity of the anonymous editor who has performed his task so skilfully. It is, however, revealed by Thomas Young (1773–1829) in what is the earliest attempt to compile a bibliography of seismology†. Young gives a list of 120 works on earthquakes, and among them occurs the entry (p. 492):

Bevis' history and philosophy of earthquakes.

John Bevis or Bevans (1693–1771) was educated at Christ Church, Oxford, where he studied medicine as a profession and optics and astronomy for pleasure. Some time before 1730, he settled in London as a physician. In 1738, he removed to Stoke Newington and there built himself an observatory in which he laboured so incessantly in taking star-transits that within seven years he had prepared his *Uranographia Britannica* or chart of the heavens in 52 plates‡. He also discovered independently the great comet of 1744. In 1750, he was elected a member of the Berlin Academy of Sciences. "He was of a mild and benevolent dis-

* London, 1757, 351 pp.

† *Lectures on Natural Philosophy*, etc. vol. 2, 1807, pp. 490–493.

‡ Owing to the failure of his publishers, this work was lost for many years. It was published, long after Bevis' death, in 1818.

position and lively temperament..., and is said to have, from modesty, concealed his authorship of several creditable works,"* including apparently *The History and Philosophy of Earthquakes*†. He must have been gratified to find three years later that the time spent on it had not been wasted. "I have taken," says Michell, "the greatest part of my authorities either from this author or the Philosophical Transactions" and *The History and Philosophy of Earthquakes* he regards as "a work well worth the perusal of those who are desirous of being acquainted with this subject."‡

4. In his choice of the ten "most considerable writers on the subject," Bevis showed much discretion. They were all then, and with perhaps two or three exceptions are still, well known. Johann Christoforus Sturmius (1635–1703) was professor of physics and mathematics at Altdorff in Germany; Martin Lister (*c.* 1638–1712), zoologist and physician, was apparently the first to suggest the construction of geological maps§; Robert Hooke (1635–1702), a man of "more than common, if not wonderful, sagacity in diving into the most hidden secrets of Nature," delivered his "discourses of earthquakes" before the Royal Society from 1667 to 1697, though, as Mallet says, they are "a sort of system of physical geology" rather than a discussion of earthquake-phenomena; John Woodward (1665–1728), the founder of the Woodwardian professorship at Cambridge, wrote *An Essay towards a Natural History of the Earth* (1695)—a work that passed through several editions and may still be read with interest; Nicolas Lemery (1645–1715) was the author of a *Cours de Chymie* that was often reprinted; Pierre Bouguer (1698–1758) spent ten years from 1735 with de la Condamine in Peru measuring a degree of the meridian near the equator; George Louis Leclerc, Comte de Buffon (1707–88), a man "of handsome person and noble presence, endowed with many of the external gifts of nature, and rejoicing in the social advantages of high

* *Dict. Nat. Biog.* vol. 4, 1885, pp. 451–452.

† A footnote on pp. 212–213, added to Buffon's text, is signed "J.B."

‡ *Phil. Trans.* vol. 51, 1761, p. 566 n.

§ Visitors to Westminster Abbey will recall the simple inscription to his little daughter "Jane Lister dear Childe."

rank and large possessions," was perhaps the most widely known of all through his encyclopaedic *Histoire Naturelle*; John Ray (1627–1705) was "the father of natural history in this country"; Stephen Hales (1677–1761) was valued both in France and England as a botanist, physiologist and inventor; and, lastly, William Stukeley (1687–1765), "a learned but honest man," won a reputation as an antiquary which has lasted to the present day*.

5. From an historical point of view, the articles collected by Bevis are useful in enabling us to estimate the knowledge of the phenomena and causes of earthquakes in the middle of the eighteenth century. The principal phenomena were clearly summarised, especially by Sturmius, Buffon and Stukeley. It was known, for instance, that, though there is hardly any country in the world that has not at one time or another been shaken by earthquakes, yet mountainous countries near the sea are exposed to the most violent, and places near volcanoes, to the most frequent, shocks; while flat, marshy, inland countries are seldom shaken, at any rate by "original" earthquakes. Buffon divides earthquakes into two classes. "One of them is occasioned by the action of subterraneous fires and explosions of volcanoes, and these are felt but to small distances, and at the time the volcanoes are raging or before their first eruption." Earthquakes of the other kind are "very different as to their effects and probably their causes too." They "are felt to vast distances and shake a long stretch of ground without the intermediation of any new

* The following are the works from which Bevis made his selection of articles:

1. Sturmius, J. C. *De Terrae-Motibus*, etc. Altdorff, 1670. 32 pp.
2. Lister, M. *Phil. Trans.* vol. 13, 1683, pp. 512–519.
3. Hooke, R. *Posthumous Works*, 1705, pp. 277–450.
4. Woodward, J. *An Essay towards a Natural History of the Earth*, etc. 3rd ed. 1723, pp. 149–160.
5. Lemery, N. Paris, *Ac. Sci. Hist. Mém.* année 1750, 1753, pp. 101–110.
6. Bouguer, P. *La Figure de la Terre*, 1749, pp. lxiv–lxxviii.
7. Buffon, G. L. L. *Histoire Naturelle*, 2nd ed. 1750, vol. 1, pp. 502–535.
8. Ray, J. *Three Physico-Theological Discourses* (second discourse, Consequences of the Deluge), 3rd ed. 1713, pp. 289–291, 294.
9. Hales, S. *Phil. Trans.* vol. 46, 1752, pp. 669–681.
10. Stukeley, W. *The philosophy of earthquakes*, a pamphlet of 139 pp. 3rd ed. 1756, based on three articles in *Phil. Trans.* vol. 46, 1752, pp. 641–645, 657–669, 731–750.

volcano or eruption." Earthquakes occur at all seasons, by night as well as by day, and under all varieties of constellations indifferently. After a very severe earthquake, such another as a rule only succeeds it after a long interval of time. The movement is sometimes a horizontal trembling, occasionally upwards (succussion). Its duration may be exceedingly short, not more than a few seconds, or may extend over whole days or "even months and years by fits." A hollow thundering noise precedes or accompanies the shock; but sometimes it is heard without any perceptible motion of the earth. In some earthquakes, torrents of water flow from fissures in the ground; in others, rivers, fountains and lakes have vanished. The surface of the earth may be raised above, or sunk below, its original level, and, in the former case, new islands may appear, though it seems probable, from the examples given, that the phenomena of earthquakes were not always kept clearly distinct from those of volcanoes.

With this accurate knowledge of earthquake-phenomena there are mingled many conceptions that can only be described as based on insufficient evidence. It is said that earthquakes are preceded and accompanied by strong winds, by fireballs and meteors, and by a continually clouded sun; that they usually occur in calm weather with a black cloud or when rain follows a great drought; and that they are succeeded by pestilences, contagious diseases and famines. Stukeley remarks that "earthquakes generally happen to great towns and cities; and more particularly to those that are situated on the sea, bays, and great rivers," or, rather more definitely as he writes in his first paper, that "the chastening rod is directed to towns and cities, where are inhabitants, the objects of its monition; not to bare cliffs and an uninhabited beach."*

6. If, with our far wider knowledge of the structure of the earth, we have much to learn as to the origin of earthquakes, it is not surprising that the solution of so complex a problem should have escaped our predecessors in the eighteenth century. They naturally connected volcanic and earthquake phenomena and

* *Phil. Trans.* vol. 46, 1752, p. 645.

assigned them to the same cause. If Woodward in 1695 could refer earthquakes to some accidental obstruction of a subterranean fire, and Stukeley to electricity, "a sort of soul to matter,"* the remaining writers, with the exception of Ray, who is silent on the question, are unanimous in favour of an explosive origin. While the accounts naturally differ in detail, the fullest is that given by Buffon.

As already noted (art. 5), Buffon divides earthquakes into two classes. The first he refers at once to volcanic explosions. As to the cause of the second kind, "it must be remembered," he says, "that all substances which are inflammable and capable of explosion do, like powder, at the instant of their inflammation, generate a great quantity of air; that air thus generated by fire is in a state of exceeding great rarefaction, and, from its circumstance of compression within the bowels of the earth, must produce most violent effects. Suppose now that at a considerable depth, as a hundred or two hundred fathoms, there should happen to be pyrites and other sulphureous matters, and that, through the fermentation excited by the filtration of waters or by any other means, they come to ignite....These matters taking flame will produce a great quantity of air, whose spring compressed in a small space, as that of a cavern, will not only shake the ground about it, but will attempt all ways of escaping and being at liberty. The passages which offer are the cavities and trenches formed by subterraneous waters and rivulets; the rarefied air will be precipitated with violence into every passage that is open to it and form a furious wind, the noise whereof will be heard on the earth's surface accompanied with shocks and tremors."†

7. It is interesting to notice how the various phenomena of earthquakes are explained on this theory. As the passages traversed by the gases are elongated, the tremors will also be

* "If a non-electric cloud discharges its contents upon any part of the earth, when in a high electrified state, an earthquake must necessarily ensue. The snap made upon the contact of many miles compass of solid earth, is that horrible uncouth noise, which we hear upon an earthquake; and the shock is the earthquake itself." (*Hist. and Phil.* p. 260.)

† *Hist. and Phil.* pp. 232–233; *Histoire Naturelle*, 2nd ed. 1750, vol. 1, pp. 528–529.

propagated lengthwise. "If this sulphureous blast be kept continually confined so as not to be able to extricate itself through any aperture, the earthquake lasts a considerable time and with strong plunges till its motion is become languid." The direction in which the shock is felt, horizontally, vertically or obliquely, depends on that of the underground passages, "just as in guns the force of the powder is directed in the same way that the piece is planted." "If an huge bulk of earth be forced up obliquely through the incumbent sea, so as not to drop back into the submarine cavern, but to rest on the solid bottom near the aperture, with its top above the surface of the sea, a new island will be formed." At the same time, much of the sea will be absorbed into the abyss below, and to this must be ascribed "the sea's instantaneous receding from the shore during an earthquake,... it being sucked into the new gulph below." Buffon, however, considers that the motion of the sea "arises solely from the shock communicated to its water by the explosion," forgetting that, in such a case, there should be a sea-wave with every strong submarine earthquake. The distribution of earthquakes is connected with that of the inflammable matter underground. Mountains are "obnoxious to shocks" because of the redundancy of inflammable substances beneath them. England is so little troubled with earthquakes and Italy so greatly because of the paucity of pyrites in the one and its abundance in the other. Lastly, according to Sturmius, "Honoratus Faber illustrates this doctrine by a variety of artificial earthquakes, as he calls them, confining gunpowder (a mixture of nitre, sulphur and charcoal) in pits, and setting fire to it by a train"*—surely the earliest of a long series of seismic experiments†.

ÉLIE BERTRAND and the *MÉMOIRES HISTORIQUES ET PHYSIQUES SUR LES TREMBLEMENS DE TERRE*

8. The same year (1757) saw the publication, not only of the *History and Philosophy of Earthquakes*, but also of a remarkable little volume by Élie Bertrand (1712–*c.* 1790), a Swiss naturalist

* Honoré, or Honorato, Fabri (*c.* 1607–88), *Physica, id est, Scientia Rerum Corpearum*, vol. 3, 1670, pp. 286–287.

† *Hist. and Phil.* pp. 26, 38, 40, 43, 62, 187, 233, 236.

and geologist, pastor at Berne, and member of the Academies of Berlin, Göttingen, Leipzig, etc.* About this time, writers on earthquakes were more intent on discussing their origin than on studying their phenomena, and Bertrand follows the prevailing custom so far as to devote more than half his volume to the "philosophy of earthquakes." Fortunately, however, he also held that the earthquakes of every country should be studied with care, and his memoirs on those observed in Switzerland possess a lasting value.

That Bertrand's book was esteemed in his own day is clear from the use that Michell makes of it in this memoir written three years later. "This author," he remarks, "in these sensible memoirs, has obliged the public with a circumstantial account of all the facts he could collect relating to the earthquakes of Switzerland, or those of other places that seemed to be connected with them. The whole seems to be done with care and fidelity, and without the least attachment to any particular system."†

9. The volume contains eight memoirs on the history of earthquakes in general and on those of Switzerland in particular. They thus fall naturally into two groups. In the more valuable section on the earthquake-phenomena of Switzerland (memoirs II–V), the longest memoir is a chronological account of the earthquakes felt in Switzerland from A.D. 563 to 1754 (pp. 22–102). The total number of earthquakes described is 155, of which 28 occurred in the sixteenth century, 62 in the seventeenth, and 31 in the eighteenth. Bertrand notices how irregularly the earthquakes are distributed over the country, the cantons most disturbed being those of Glaris, Basel and Bern, and portions of Valais, Vaud and Zurich. The fourth memoir (pp. 117–142) is one of our earliest detailed accounts of a strong earthquake—that of 9 Dec. 1755; while the fifth (pp. 143–168) deals with earthquakes observed in the Haut-Valais in 1755–56. In the third memoir (pp. 103–116) are collected the observations on the effects of the Lisbon earthquake in Switzerland—the seiches

* *Mémoires Historiques et Physiques sur les Tremblemens de Terre*, La Haye, 1757, 328 pp.

† *Phil. Trans.* vol. 51, 1761, p. 568 n.

produced in the lakes and the disturbance of many springs. It is a useful companion to the last chapter of *The History and Philosophy of Earthquakes.*

10. There is little need to refer in detail to Bertrand's clearly expressed views on the cause of earthquakes (pp. 1–21, 169–237, 300–326), for they are almost the same as those of Buffon described above (art. 5). Pyrites and pyritous materials occur, in greater or less quantity, in every place. When moistened, they become warm, ferment, and sometimes even take fire. The interior air thus dilated but confined in subterranean channels and caverns impinges on any obstacles that obstruct its free dilatation, and so gives rise to earthquakes. And the reason why certain parts of Switzerland are more subject than others to earthquakes is that they are more cavernous and contain more mineral springs and beds of sulphur. This Bertrand considers the most probable explanation, but he admits that it is by no means a complete one. There are phenomena that it is difficult so to explain—such, for instance, as the great extent of the area disturbed by the Lisbon and other destructive earthquakes, the high velocity with which the movement travels, and the regularity in the nature and direction of the undulations. Every movement due to a fermentation or a sudden inflammation must, he realises, be communicated successively, and must be confused, tumultuous, without order or direction.

11. The phenomena of earthquakes—the touchstone of his theory—are fully described by Bertrand (pp. 238–299), more fully than by Sturmius and Stukeley (art. 5). He notices that earthquakes often occur closely together, sometimes only a few minutes or even a few seconds apart, and that these returns are governed by no periodic law. The sound frequently precedes the shock, but by an interval too brief to act as a prognostic. That which accompanies the shock he likens to the fall of a load, an explosion such as that of a cannon, a rolling like that of thunder. The noises, he holds, are the effects of different causes, one of them being the collision of the solid parts of the interior. Bertrand was one of the first, if not the first, to realise the rapid propagation of the earthquake-motion. With regard to the time of occurrence, he was not in advance of his contemporaries. Earthquakes, he remarks, are more frequent in spring and autumn than in winter and summer, and during the night than in the daytime. They avoid times of great heats or great colds, but, as noticed above (art. 5), they often occur when heavy rains succeed a long period of drought.

Lastly, Bertrand notices that earthquakes travel without hindrance across plains, under mountains, and below the deepest valleys. Surface-features, indeed, have no effect whatever in modifying or directing the propagation of the movement. And these are evident proofs, he urges, that the "active principle" lies not just under, but at a great depth below, the surface of the earth.

What is remarkable about the theoretical half of Bertrand's book is the skill with which the theory is presented, the author's ability to see its weak, as well as its strong, points, the justice of much of his criticism, and the independence of his judgment. At the present time, we should be inclined to reverse the order which he assigned to his two parts, and to give the prior place to the history of the "earthquakes of Switzerland in particular" rather than to that of "earthquakes in general." His catalogue must certainly rank as the earliest attempt, and one of the best attempts, to construct the seismic chronicle of a district in which earthquakes are frequent and often strong, though never of very destructive violence.

CHAPTER II

JOHN MICHELL

12. The date and place of Michell's birth are unknown, but the year must have been 1724 or 1725 and the place may have been Nottingham. Coming at any rate from that city, he entered Queens' College, Cambridge, in 1742, and graduated as fourth wrangler in the mathematical tripos of 1748–49. This success which, but for his wide interests, would probably have been greater, was followed by his election to a fellowship at Queens'. His wonderful versatility, even at so early a stage, is shown by his ability to give college lectures on Greek and Hebrew as well as on mathematics. At this time, too, he must have been making those original observations on the structure of the earth's crust, which are described in his memoirs on earthquakes. Two years later, he was appointed Woodwardian professor of geology, being the third holder of that office and the most capable until Sedgwick began his long reign in 1818. In addition to the college tutorship and lectures—there is no evidence that the duties of the professorship were a serious burden—Michell held the rectory of St Botolph's, Cambridge, from 1760 to 1763. The latter year, however, brought a change, the busy official life at Cambridge giving place to the studious quiet of a country rectory, first at Compton, near Winchester, then at Havant, and lastly, in 1767, at Thornhill, a Yorkshire village not far from Dewsbury. The fellowship at Queens' and the professorship of geology were both vacated in 1764, the year of his first marriage and as a result of that step. In a smaller man such a life might have ended in stagnation, but, with his musical evenings at home, his friendship with many leading men of science, and his frequent visits to London, Michell maintained his freshness, and science gained, rather than lost, by the change. Happy in his social life, unfailing in the discharge of his parochial duties, and evidently possessing means sufficient for his wants, Michell carried on his scientific work without pause and without hurry, touching nothing that he did not adorn, and careless, apparently, to whom the credit might fall so long as the work was done. He died at Thornhill on 21 Apr. 1793 in his 69th year*.

13. Except for his early *Treatise of Artificial Magnets* and his memoir on earthquakes, Michell's work was mainly astronomical. At the age of 25, he discovered the method of making magnets by double touch and the variation of magnetic action according to the inverse square of the distance. Less than ten years later, he developed those clear views on stratification that are fully appreciated by modern geologists†. Equally original and no less important was his memoir on stellar parallax (1767). In this he insisted on the extreme minuteness of the parallax of even the brightest stars, and inferred that the nearest fixed star is probably no farther from us than 220,000 times the sun's distance—a quite close, if accidental, approximation to the distance of α Centauri. The apparent proper motion of the stars he correctly attributed in part to that of the sun; he estimated the extreme probability that the stars which form the Pleiades constitute a connected system, and he foresaw the discovery, made shortly afterwards by William Herschel (1738–1822), of the revolutions of the double stars. At Thornhill, he experimented on the best alloys for the mirrors of reflecting telescopes and on the method of giving the mirrors their proper form. The 10-feet reflector that he made was afterwards bought and used by Herschel. Michell also invented the torsion-balance, and devised a method

* *Dict. Nat. Biog.* vol. 37, 1894, pp. 333–334; *Eng. Mechanic*, vol. 13, 1871, pp. 309–310; *Knowledge*, vol. 15, 1892, pp. 188–191, 206–208; Sir A. Geikie, *Memoir of John Michell*, 1918.

† Though they are not connected with the cause of earthquakes, Michell's remarkable views on mountain-structure are worth quoting. Mountains are "generally," he says, "if not always, formed out of the lower strata of the earth. This situation of the strata may be not unaptly represented in the following manner. Let a number of leaves of paper, of several different sorts or colours, be pasted upon one another; then bending them up together into a ridge in the middle, conceive them to be reduced again to a level surface, by a plane so passing through them, as to cut off all the part that had been raised; let the middle now be again raised a little, and this will be a good general representation of most, if not all, large tracts of mountainous countries." In some instances, he adds, "it often happens, that the hills, to which these high lands serve as a base, are not only formed out of the strata next above them, but they stand, as it were, in a dish, as if they had depressed the ground, on which they rest, by their weight."

of determining the density of the earth. The apparatus was completed a short time before his death, but he did not live to make any experiment with it. It passed ultimately into the capable hands of Henry Cavendish (1731–1810), and led to his well-known determination of the density of the earth. Michell, like Darwin, was evidently one of those men who "seemed by gentle persuasion to have penetrated that reserve of nature which baffles smaller men." Is it too much to claim for him a place, not, indeed, beside Newton and Darwin, but in a rank not so very far below them?

During his own life, Michell's extraordinary power was known and valued. Among his friends he could number such men as Cavendish, Black (1728–99), Priestley (1733–1804) and William Herschel. It is all the stranger, then, that his memoir on the cause and phenomena of earthquakes should have remained, or appeared to remain, practically unknown for more than half a century. It was referred to in terms of strong praise in the *Edinburgh Review* for February 1818 (p. 318), and later in that year was reprinted, with notes by John Farey (1766–1826), in the *Philosophical Magazine**. Many years after his death, full, if belated, justice has been done to Michell's stratigraphical conceptions; but his work, so far as it deals with earthquakes, is less well known, owing partly perhaps to the inaccurate account that Mallet gave of it in his early memoirs†.

14. In one respect, Michell differs noticeably from his predecessors—he freed himself entirely from the shackles of the old classical writers and relied only on the evidence of modern observers. This welcome change, however, was not due altogether to him. Taking Bevis' *History and Philosophy of Earthquakes* as a whole, two out of every five references are to the works of Aristotle, Seneca, Pliny and other writers of a time long past.

* *Conjectures concerning the cause and observations upon the phenomena of earthquakes: particularly of that great earthquake of the first of November*, 1755, *which proved so fatal to the city of Lisbon, and whose effects were felt as far as Africa, and more or less throughout all Europe* (read 28 Feb., 6, 13, 20, 27 Mar. 1760). *Phil. Trans.* vol. 51, 1761, pp. 566–634; *Phil. Mag.* vol. 52, 1818, pp. 186–195, 254–270, 323–340.

† *Irish Ac. Trans.* (read 9 Feb. 1846), vol. 21, 1848, pp. 58–60, 65–67, 84–85; *Brit. Ass. Rep.* 1850, pp. 17–19.

In the first article, that by J. C. Sturmius, less than half the references (46 per cent.) are to comparatively modern writers. In Bertrand's *Mémoires Historiques et Physiques*, three out of every four references are to recently published works. Throughout the whole of his memoir, Michell refers but once to classical writers, and then only in a quotation from Bertrand, and the earthquakes on which the memoir is based were all at the time more or less recent, the earliest being the Jamaica earthquakes of 1687–88 and 1692, and the latest the Lisbon and New England earthquakes of 1755.

15. Leaving out of account the important section on the nature and trend of strata—a section which has been adequately dealt with by Sir A. Geikie and others—Michell's paper may be divided into three parts. In the first he describes the phenomena of earthquakes, in the second he explains in detail his theory of their cause, and in the third he gives methods for determining the position of the seismic focus.

16. The Phenomena of Earthquakes. Like his predecessors, Michell begins with a summary of the phenomena which his theory must explain; but, unlike them, he distinguishes between certain phenomena which are essential to, and others which have no connexion with, the origin of earthquakes. The latter are what we now call secondary effects of the shock, such as the sudden stopping or gushing out of fountains occasioned by the opening or contraction of fissures, the dizziness and sickness induced by the wave-like motion, and the disturbances sometimes occasioned in the direction of the magnetic needle. These are not "observed to be constant attendants on earthquakes, nor do they seem materially to affect the solution given one way or other." The phenomena which Michell regards as essential, five in number, are the following:

(i) The same places are subject to returns of earthquakes, not only at small intervals for some time after any considerable one has happened, but also at greater intervals of some ages. Michell thus distinguishes between the occurrence of after-shocks and the returns of great earthquakes, as he shows by reference to the 451 shocks felt at Lima from 28 Oct. 1746 to 24 Feb. 1747, and to the sixteen very violent earthquakes at the same place from 1582 to 1746.

(ii) Those places that are in the neighbourhood of burning mountains are always subject to frequent earthquakes, and the eruptions of those mountains, when violent, are generally attended by them. Michell refers especially to Chili and Peru, no other known countries, he supposes, being so subject to earthquakes and none so full of volcanoes.

(iii) The motion of the earth in earthquakes is partly tremulous and partly propagated by waves, which succeed one another sometimes at larger, sometimes at smaller, distances; and this latter motion is usually propagated much farther than the other. The former part of this statement, Michell considers, "wants no confirmation." Indeed, his explanation of the manner in which the vibrations are propagated through the solid crust is quite as clear as that given by Mallet nearly a century later. In illustration of the latter, he quotes accounts of what are now called "visible waves," such as that of an observer of the Jamaica earthquake of 1688, who "saw the ground rise like the sea in a wave as the earthquake passed along, and" who "could distinguish the effects of it to some miles distance by the motion of the tops of the trees on the hills." In the Lisbon earthquake, "this wave-like motion was propagated to far greater distances than the other tremulous one, being perceived by the motion of waters and the hanging branches in churches through all Germany... in Denmark, Sweden, Norway, and all over the British Isles."

(iv) It is observed in places which are subject to frequent earthquakes that they generally come to one and the same place from the same point of the compass. Also, the velocity with which they proceed (so far as one can collect it from the accounts of them) is the same; but the velocity of the earthquakes of different countries varies. Michell's illustration of the latter statement, though not quite accurate, is interesting. The velocity with which the Lisbon earthquake and its after-shocks were propagated "was the same, being at least equal to that of sound; for all followed immediately after the noise that preceded them, or rather the noise and the earthquake came together"; that of the New England earthquakes was less, as is shown by the longer interval between the preceding noise and the shock*.

(v) The great Lisbon earthquake was succeeded by several local ones (in Switzerland and elsewhere), the extent of which was much less.

17. The Cause of Earthquakes. Long before Michell's time, it was "the general opinion of philosophers that earthquakes owe their origin to some sudden explosion in the internal parts of the earth." Michell traces earthquakes to the primary cause to which they also appeal, namely subterraneous fires. He differs from his predecessors chiefly in his endeavour to support this theory by facts and to trace out the effects of such explosions. "These fires," he says, "if a large quantity of water should be let out upon them suddenly, may produce a vapour whose quantity and elastic force may be fully sufficient for the purpose." Indeed, he remarks, "it is not easy to find any other cause capable of producing such sudden and violent effects."

The fires to which he refers are of the same kind as the fires of volcanoes, but, like Woodward before him and Humboldt after him†, Michell regards the volcanic fires as safety-valves, "for volcanoes, giving passage to the vapours that are thus formed, should rather prevent" earthquakes. If it should be asked why we should suppose that subterraneous fires exist in the neighbourhood of volcanoes, Michell points to the frequent instances of new volcanoes breaking out near old ones, to the existence of many volcanoes close together in the same tract of country, and to the rarity of isolated volcanoes.

Michell regards it as very probable that the fires of volcanic regions originate in a stratum, like coal or shale, in which pyrites is lodged in such quantity that it will inflame of itself, but that the quantity varies in different parts of the stratum so that the fires are not continuous. The same strata, he remarks, are generally very extensive, and they commonly lie more inclining

* From records of the time of the great earthquake at Lisbon and more distant places, Michell estimates the velocity at more than 20 miles a minute. This is the earliest estimate of the kind known to me.

† John Woodward, *An Essay toward a Natural History of the Earth*, 1695, p. 144, and art. 45.

from the mountainous countries than the countries themselves.

"These circumstances make it very probable that these strata of combustible materials which break out in volcanoes at the tops of the hills are to be found at a considerable depth under ground in the level and low countries near them. If this should be the case, and if the same strata should be on fire in any places under such conditions, as well as on the tops of hills, all vapours, of whatsoever kind, raised from these fires must be pent up, unless so far as they can open for themselves a passage between the strata; whereas the vapours raised from volcanoes find a vent, and are discharged in blasts from the mouth of them. Now, if when they find such a vent they are yet capable of shaking the country to the distance of ten or twenty miles round, what may we expect from them when they are confined?" Thus, "the most extensive earthquakes should take their rise from the level and low countries, but more especially from the sea, which is nothing else than waters covering such countries."

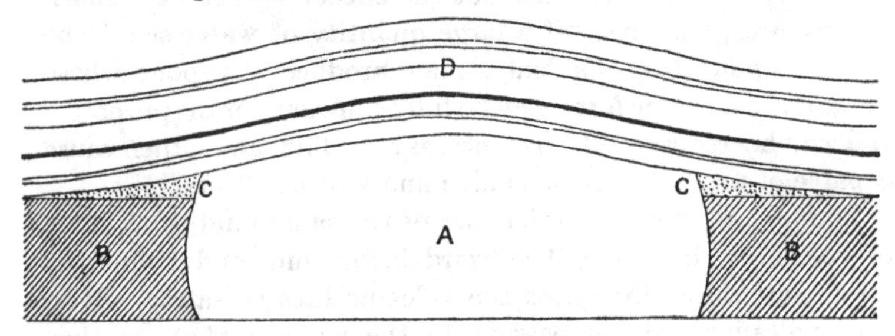

Fig. 1. Diagram illustrating Michell's theory of the origin of earthquakes.

The manner in which the water contained in fissures reaches the subterraneous fires below is explained at some length. Michell supposes the roof over a subterraneous fire to fall in. "If this should be the case, the earth, stones, etc., of which it is composed, would immediately sink in the melted matter of the fire below; hence all the water contained in the fissures and cavities of the part falling in would come in contact with the fire and be almost instantly raised in vapour."

The fall of the roof, he realises, could not happen accidentally, and he illustrates the way in which it might occur by the accompanying diagram (Fig. 1). In this, *A* is supposed to represent a vertical section of the matter on fire; *BB* parts of the same stratum yet unkindled; *D* the strata lying over the fire, which are raised a little and bent by the expansion of the heated matter at A; and CC the annular space so formed around the fire*.

This space will be gradually filled with water, as it is formed, the melted matter being prevented from filling it, by its want of fluidity...; for the lentor and sluggishness of this kind of matter is such, that, when somewhat cooled on the surface by the contact of the air only, it will not flow, perhaps, ten feet in a month, though in a very large body; instances of which we have in the lavas of Ætna, Vesuvius, etc. It is not to be expected then, that it should spread far, when it comes in contact with water at its edges, as soon as it is formed, and when it is, perhaps, several months in acquiring a thickness of a few inches; but it must, by degrees, form a kind of wall between the fire and the opening into the annular space....This wall will gradually increase in height, till it becomes too tall in proportion to its thickness, to bear any longer the pressure of the melted matter; which must necessarily happen at last, because the thickness of it will not exceed a certain limit.

18. Michell then proceeds to show how his theory will account for the various phenomena described above. The return of earthquakes to the same place is to be expected, for, as in volcanoes, the subterraneous fires may be supposed to subsist for many ages. Moreover, as volcanoes frequently rage for a time and then are quiet for a number of years, so we see earthquakes frequently repeated for some small time and then ceasing for a long period, except perhaps for the occurrence now and then of some slight shock. The great frequency of earthquakes in the neighbourhood of burning mountains is a strong argument for their proceeding from a cause of the same kind.

The two forms of motion which Michell so clearly discerned in earthquakes are consequences, in his view, of the varying amount of the vapour generated. A small quantity of vapour almost instantly generated would produce the vibratory motion; a very large quantity would result in a wave-like motion.

The first effect of the vapour, he says, would be to form a cavity between the melted matter and superincumbent earth by the compression of the materials immediately over the cavity, and "this compression must be propagated on account of the elasticity of the earth, in the same manner as a pulse is propa-

* To give precision to his explanation, Michell suggests that the horizontal extent of the fire at A may be from $\frac{1}{2}$ mile to 10 or 20 miles, the thickness of the stratum B to be from 10 or 20 to 100 yards, the height of the annular space C next the fire to be from 4 or 5 to 10 or 15 feet, and the thickness of the superincumbent matter at D to be from $\frac{1}{4}$ or $\frac{1}{2}$ mile to 2 or 3 miles.

gated through the air; and again the materials immediately over the cavity, restoring themselves beyond their natural bounds, a dilatation will succeed to the compression; and these two following each other alternately for some time a vibratory motion will be produced at the surface of the earth. If these alternate dilatations and compressions should succeed one another at very short intervals, they would excite a like motion in the air, and thereby occasion a considerable noise. The noise," he adds, "that is usually observed to precede or accompany earthquakes is probably owing partly to this cause, and partly to the grating of the parts of the earth together, occasioned by that wave-like motion before mentioned."

19. Michell illustrates the propagation of the wave-like motion by describing an experiment which has led to some misconception of his views.

Suppose a large cloth or carpet (spread upon a floor) to be raised at one edge and then suddenly brought down again to the floor, the air under it, being by this means propelled, will pass along until it escapes at the opposite edge, raising the cloth in a wave all the way as it goes. In like manner, a large quantity of vapour may be conceived to raise the earth in a wave, as it passes along between the strata, which it may easily separate in an horizontal direction, there being...little or no cohesion between one stratum and another. The part of the earth that is first raised, being bent from its natural form, will endeavour to restore itself by its elasticity, and the parts next to it, beginning to have their weight supported by the vapour which will insinuate itself under them, will be raised in their turn, till it either finds some vent or is again condensed by the cold into water, and by that means prevented from proceeding any further.

Equally interesting and less open to objection is Michell's explanation of the propagation of the wave-like motion to distances far greater than the vibratory motion. This motion, he states, "will be propagated through the solid parts of the earth, and, therefore, it will much sooner become too weak to be perceived than the wave-like motion, for this latter, being occasioned by the vapour insinuating itself between the strata, may be propagated to very great distances; and even after it has ceased to be perceived by the senses it may still discover itself by the appearances before mentioned." That is to say, in modern terms, the wave-like motion diverges in two, and the vibratory motion in three, dimensions.

All earthquakes derived from the same subterraneous fire must come to any place from the same direction. Those only which are derived from different fires will come from different parts of the compass. Moreover, since, according to Michell, the velocity with which the vapour insinuates itself between the strata depends upon its depth below the surface, earthquakes must travel with the same velocity if they come from the same place, and with different velocities if they proceed from fires at different depths. The occurrence of local earthquakes in other places, Michell accounts for by supposing that the vapour coming from one fire may disturb the roof over some adjacent fire, and so occasion earthquakes by the falling in of some part of it.

20. Michell also suggests explanations of two other phenomena in addition to those described above. These are the great sea-waves and "a sort of periodical return" of earthquakes in the same region. The first movement of the sea at Lisbon was, as usual, a retreat of the water followed by the inrush of the great waves. During the elevation of the roof over the subterraneous fire, the waters of the ocean lying over it must retreat, but so slowly and so gently as to occasion no great disturbance. But, as soon as some part of the roof falls in,

> The cold water contained in the fissures of it, mixing with the steam, will immediately produce a vacuum, in the same manner as the water injected into the cylinder of a steam-engine, and the earth subsiding and leaving a hollow place above, the waters will flow every way towards it, and cause a retreat of the sea on all the shores round about; then presently, the waters begin again converted by the contact of the fire into vapour,...the earth will be raised, and the waters over it will be made to flow every way, and produce a great wave immediately succeeding the previous retreat.

21. Lastly, the great quantity of water let out upon the subterraneous fire "must extinguish a great portion of the burning matter, in consequence of which, it will be contracted within much narrower bounds," and this will continue "till the roof is well settled, and the surface of the melted matter sufficiently cooled, after which, it may require a long time for the fire to heat it again so much, as will be necessary to make it produce the former effects."*

* At the end of the section on stratification, Michell gives the following paragraph: "Besides the rising of the strata in a ridge, there is another very remarkable appearance in the structure of the earth, though a very common

22. The Position of the Seismic Focus. In the last section of his memoir, Michell shows that,

If we would inquire into the place of origin of any particular earthquake, we have the following grounds to go upon.

(i) The different directions, in which it arrives at several different places: if lines be drawn in these directions, the place of their common intersection must be nearly the place sought; but this is liable to great difficulties; for there must necessarily be great uncertainty in observations, which cannot, at best, be made with any great precision, and which are generally made by minds too little at ease to be nice observers of what passes, moreover, the directions themselves may be somewhat varied, by the inequalities in the weight of the superincumbent matter, under which the vapour passes, as well as by other causes.

(ii) We may form some judgement concerning the place of the origin of a particular earthquake from the time of its arrival at different places; but this also is liable to great difficulties. In both these methods, however, we may come to a much greater degree of exactness, by taking a medium amongst a variety of accounts, as they are related by different observers. But,

(iii) We may come to the greatest degree of exactness in those cases, where earthquakes have their source from under the ocean; for, in these instances, the proportional distance of places from that source may be very nearly ascertained, by the interval between the earthquake and the succeeding wave; and this is the more to be depended on, as people are much less likely to be mistaken in determining the time between two events, which follow one another at a small interval, than in observing the precise time of the happening of some single event.

The first method is that which Mallet applied with such skill in his study of the Neapolitan earthquake of 1857 (arts. 72–76), though he depended on more permanent evidences of the direction than the impressions of observers. The third method is of course vitiated by the assumption that the velocity of sea-waves is independent of the depth of the ocean; but the greater accuracy attainable by measuring intervals rather than absolute times lies at the root of the method depending on the duration of the preliminary tremor which is now so widely used.

In locating the epicentre of the Lisbon earthquake, Michell relies on other evidence rather than on the methods which he

one; and this is what is usually called by miners the trapping down of the strata; that is, the whole set of strata on one side a cleft are sunk below the level of the corresponding strata on the other side. If, in some cases, this difference in the level of the strata on the different sides of the cleft should be very considerable, it may have a great effect in producing some of the singularities of particular earthquakes." One cannot help regretting the brevity of Michell's statement on the relations between earthquakes and faults. I can only suggest that, according to his view, the vapour, in travelling outwards by parting adjacent strata, would be suddenly arrested at the fault.

describes so clearly. The great sea-wave which followed the earthquake indicates that its origin was submarine, and it could not be far from Lisbon, as the wave arrived there so soon after the earthquake and was of such great height. It must, he thinks, lie somewhere between the latitudes of Lisbon and Oporto, though probably somewhat nearer the former (about a degree of a great circle, he afterwards assumes, from Lisbon and a degree and a half from Oporto), and at a distance of 10 or 15 leagues from the coast. Such a position agrees well with the few observations which he possessed on the direction and time of the earthquake. But, curiously enough, he notices that "the times which the wave took up in travelling are not in the same proportion with the distances of the respective places from the supposed source of the motion." He does not, however, regard this as an objection to the point assumed, but thinks—and here he foresaw a later discovery—that "the true reason of this disproportion seems to be the difference in the depth of the water; for, in every instance in the above table, the time will be found to be proportionately shorter or longer as the water through which the wave passed was deeper or shallower."

23. The method that Michell suggests for inquiring "into the depth at which the cause lies that occasions any particular earthquake" depends on his theory of the origin of earthquakes and need not be considered here. For the Lisbon earthquake, the observations required in order to apply the methods did not exist. But "if," he says, "I might be allowed to form a random guess about it, I should suppose (upon a comparison of all circumstances) that it could not be much less than a mile or a mile and a half, and I think it is probable it did not exceed three miles."*

24. Conclusion. In this last section, I propose to give a brief summary of Michell's contributions which seem to me to possess a permanent value.

(i) In the first place must be mentioned his distinction between the phenomena which are and are not essential to his

* From the footnote on p. 605 of his memoir, it is clear that Michell imagined that the depth of the focus might be even a quarter or half a mile (art. 17).

theory. In separating the vibratory motion from the wave-like motion or visible waves, Michell was in advance of his time. He was one of the first, if not the very first, to assign the vibratory motion in earthquakes to the propagation of elastic waves in the earth's crust.

(ii) In the second place he rendered an important service in his attempt to give a consistent theory of the origin of earthquakes, even if the theory, as we can now see, rests on an erroneous conception of volcanic action, and if it cannot explain all the phenomena that have since been discovered. His suggestions as to the origin of earthquake-sounds and of seismic sea-waves would seem, in part at any rate, to be correct.

(iii) Lastly, Michell devised two useful methods for determining the position of the epicentre. Both were afterwards overlooked, and both have been re-invented and applied with varying success to the investigation of earthquakes. Michell's "random guess" as to the depth of the focus, like Newton's guess with regard to the density of the earth, is one of those intuitions which only occur to the ablest minds.

Michell evidently failed to impress his readers, and "he who succeeds in doing so," says Darwin, "deserves, in my opinion, all the credit." But what if the failure be the fault of his readers —not of his contemporaries, for it was chiefly to this memoir that Michell owed his election to the Royal Society, but of their successors. If Michell's conceptions of the form and trend of strata could sink into temporary oblivion, it is not surprising that his almost prophetic insight into earthquake-phenomena should share their fate. But, if his influence on future workers were less than his memoir deserves, he does not, I think, forfeit his claim to be regarded as one of the Founders of Seismology.

CHAPTER III

FROM MICHELL TO PERREY

25. Michell's great paper was followed by a long pause in the history of seismology. From 1760 to 1845*, there was no one who stands out prominently as a founder, none to whom the study of earthquakes was the main work of his life. But, if we compare our knowledge of earthquake-phenomena in 1840 or 1845 with that in 1760, we shall realise that the advance, if made slowly, was not inconsiderable. At first, this was due to the occurrence of some remarkable earthquakes—the Calabrian earthquakes of 1783, the New Madrid earthquakes of 1811–12, the Kutch earthquake of 1819, the Valparaiso earthquake of 1822 and the Concepcion earthquake of 1835. The closing years of the period, say from 1821 onwards, were marked by a growing interest in earthquake-phenomena generally. Hoff began his yearly lists of earthquakes in 1821, and published his catalogue of earthquakes in the whole world in 1840 and 1841. The first edition of Lyell's *Principles of Geology*, with its well-known chapters on earthquakes, appeared in 1830, Humboldt's *Personal Narrative* (1811–29) contained accounts of several South American earthquakes, though his *Cosmos* (1845–58) was not published until the next period (that of Perrey and Mallet) had advanced by several years. David Milne, afterwards Milne Home, studied the earthquakes of Comrie in Perthshire from 1839 to 1844, and compiled the most detailed catalogue of British earthquakes published up to that time.

THE CALABRIAN EARTHQUAKES OF 1783

26. Very different from the Lisbon earthquake of 1755 were the earthquakes that ruined Calabria in 1783—the one, omitting after-shocks, a single great earthquake felt over a million or

* Perrey's work on earthquakes began in 1841 and Mallet's in 1846, but the work done by others from 1841 to 1845 belongs rather to the period from 1760 to 1840.

more square miles; the others, six disastrous shocks, not one of which disturbed an area of 100,000 square miles. The Calabrian earthquakes were of consequence, not so much from the destruction wrought by them or from the loss of life—at least 35,000 persons were killed—as from the fact that there were capable investigators at hand and careful recorders of the after-shocks. Few earthquakes have contributed more to the progress of seismology than those which brought havoc to Calabria Ulteriore in the early months of 1783.

The first to study the ruined country, the first also to suggest methods of relief, was F. A. Grimaldi (1740–84), the Neapolitan secretary of war. G. Vivenzio, court physician, had access to the official reports of those who brought help to the sufferers. Early in April, that is, soon after the last of the great earthquakes, a commission appointed by the Reale Accademia delle Scienze e delle Belle Lettere di Napoli went carefully over the whole of the central region. Sir William Hamilton (1730–1803), British envoy extraordinary at the court of Naples, also crossed the same district, followed soon after by the French geologist, Déodat de Dolomieu (1750–1801). No less useful, though perhaps less imposing, was the part of those who kept chronicles of the after-shocks—of Andrea de Leone at Catanzaro, of Andrea Gallo at Messina, of Girolamo Minasi at Scilla, and especially of Domenico Pignataro (1735–1802) at Monteleone*.

27. Of the Italian investigators, Francesco Antonio Grimaldi (1740–84), though hardly then attained to middle age, was widely known as the author of a life of Diogenes, of reflections on the natural inequality of men, and, especially, of six volumes of the great *Annali del Regno di Napoli*. In 1783, he was secretary of war. Trampling on private sorrow†, and in weak health, he obeyed the wish of his king and visited many of the ruined towns and villages. He died in the following year before his report on the earthquakes was published‡.

* These lists were afterwards used by Alexis Perrey in illustration of his third law of earthquake-frequency (art. 59).

† He had lost six members of his family through the earthquake.

‡ M. Delfico, *Eulogio del Marchese D. F. A. Grimaldi*, Naples, 1784, 55 pp.; *Biografia degli Italiani Illustri*, vol. 7, 1840, pp. 94–97.

This report is a small volume of 87 pages*. Grimaldi distinguished five of the earthquakes as far more violent than the others†, and regarded the earthquake of 28 Mar. as the strongest of the series. That the resulting loss of life was less then than on 5 Feb. was due, he thought, to the fact that the inhabitants were no longer living in apparent security indoors. The principal section (pp. 4–25) records in detail the destruction wrought by the earthquake of 5 Feb. and the number of lives lost at each place. The earthquake-chronicle is continued up to 12 July 1783. Grimaldi notices that the violent earthquakes affected different districts, and he refers to the official measures of relief. He also gives a useful list of destructive earthquakes in Calabria from 1181 to 1756.

28. The history of the Calabrian earthquakes by Giovanni Vivenzio, chief physician in the court of Naples, is the first of a long series of monographs on great earthquakes‡. In its final form, it is also one of the largest, for its two quarto volumes contain 569 pages and 21 plates. Much of it, however, is irrelevant so far as the earthquakes themselves are concerned. The first part (pp. 1–78) of Vol. 1, for instance, reprints a memoir by Bertholon on earthquakes from the earliest times and on a theory, warmly endorsed by Vivenzio, that earthquakes are electrical phenomena. The text of this memoir is "enriched" by editorial notes so copious that they amount to four times the length of the memoir. Then comes an equally long account (pp. 79–159) of the physical, political and ecclesiastical condition

* *Descrizione de' Tremuoti accaduti nelle Calabrie nel* 1783, Naples, 1784.

† There were in reality six great earthquakes—(i) 5 Feb., 0.45 p.m., in the Palmi zone, (ii) 6 Feb., about 1.6 a.m., in the Scilla zone, (iii) 7 Feb., 8.20 p.m., in the Monteleone zone, (iv) 7 Feb., about 10 p.m., in the Messina zone, (v) 1 Mar., 8.30 a.m., in the Monteleone zone, and (vi) 28 Mar., 1.16 a.m., in the Girifalco zone. The third and fourth were regarded as one by the early investigators.

‡ *Istoria de' tremuoti avvenuti nella Provincia della Calabria ulteriore, e nella Città di Messina nell' anno* 1783, etc., Naples, 1788. This is the second and much enlarged edition of an earlier work—*Istoria e teoria de' tremuoti in generale e in particolare di quello della Calabria e di Messina del* 1783, Naples, 1783, 446 pp. It is this latter work that is criticised by Dolomieu (p. 274, English translation of his memoir) as containing few facts interesting to the naturalist and apparently written to support the theory that ascribes the origin of earthquakes to electricity.

of Calabria Ulteriore. Thus, one-quarter of the history deals with preliminary matter. The earthquakes themselves are dismissed in nine pages (pp. 159–168), and the second part concludes with a detailed report (pp. 168–274) on the damage in the various towns and villages. In the third part (pp. 275–427) are described the measures adopted to relieve the sufferings of the survivors and, in some detail, the lakes (215 in number) formed in one way or another by the earthquakes. It is thus evident that Vivenzio regarded the earthquakes not as incidents in the growth of the earth so much as deplorable events in the history of the kingdom of Naples, and this gives his report a political tone, though in it are to be found many facts of scientific interest.

29. Not the least useful part of Vivenzio's report is the series of tables which form the text of the second volume. Of these, by far the most valuable is the "Giornale Tremuotico," compiled by Domenico Pignataro (1735–1802), a physician of Monteleone (pp. i–lxxxiii). The interval covered by the list extends from 1 Jan. 1783 to 1 Oct. 1786. Not only does he give the times of 1186 shocks, but he assigns to each its intensity in terms of a scale which, though rough and containing only five degrees, is of interest as the earliest known attempt to construct an intensity scale. He classifies the shocks as slight, moderate, strong, and very strong, denoting these intensities in his table by the symbols F', F'', F''' and F''''. In addition to these, he places the five most violent earthquakes* in a separate class, denoting them in the table by a Maltese cross. The numbers of shocks under each degree are as follows:

Year	Slight	Moderate	Strong	Very strong	Violent	Total
1783	503	235	175	32	5	950
1784	91	34	16	3	—	144
1785	27	17	4	2	—	50
1786 (to Oct. 1)	21	17	3	1	—	42
Total	642	303	198	38	5	1186

The next table (by Vivenzio) is also of value—the "Indice generale de' paese di Calabria ulteriore." It gives the condition

* Pignataro agrees with Grimaldi, Sarconi and others in regarding the two great earthquakes on 7 Feb. as one and the same.

of each town and village after the earthquakes, the numbers of their inhabitants before the first earthquake, and the numbers of persons killed. The scientific interest of the table lies in the fact that the varying percentages of deaths provide some indication of the distribution of intensity throughout the province. Some of the percentages are unusually high, for instance, 48·0 at S. Giorgio, 50·0 at Polistena, 50·9 at Oppido, 57·1 at Bagnara, and 76·9 at Terranova. The table ends with a summary, in which the different amounts of damage are classified, forming a table not unlike that drawn up by M. Baratta to represent the distribution of intensity within the meizoseismal area of the Messina earthquake of 1908*. Thus, Vivenzio remarks that there were 32 places entirely destroyed to be rebuilt elsewhere, 148 entirely destroyed to be rebuilt on the same sites, 14 totally damaged and uninhabitable, 91 in part destroyed and in part rendered uninhabitable, 44 in part destroyed and in part damaged, 26 only damaged, and so on. Unconsciously, no doubt, Vivenzio has thus drawn up a scale of intensity for destructive shocks.

30. On the same scale as the first edition of Vivenzio's *Istoria*, and published before the second and larger edition of that work, is the detailed report issued by the Neapolitan Academy of Sciences and Fine Letters†. The Commission nominated by the Academy is of interest as the first appointed to undertake the investigation of a great earthquake. It consisted of five members under the direction of Michele Sarconi (1731–97), who, at an early age, had earned fame in the medical circles of Europe by his history of the plague of 1764. As perpetual secretary of the Academy, and also as a man of wide knowledge, he was a fitting leader of this important enterprise, and it fell of course to his lot to compile the great report published in the following year. This work being finished, Sarconi devoted himself once more to his medical studies. He died in 1797 from

* *La Catastrofe Sismica Calabro-Messinese* (28 *dicembre* 1908), 1910, pp. 214–215.

† *Istoria de' fenomeni del tremoto avvenuto nelle Calabrie e del Valdemone nell' anno* 1783, Naples, 1784.

fever caught while endeavouring to save the life of a stricken friend*.

Starting early in April 1783 and separating into three sections, each accompanied by one or more artists, the Commission examined altogether more than 150 towns and villages, all damaged, and many completely destroyed, by the earthquakes. Their report is a massive volume of 372 folio pages, illustrated by 69 plates and 9 double-page maps. By far the larger part (pp. 1–320) is occupied by descriptions of the damage in each of the places visited, with occasional remarks on the after-shocks and sea-waves. It is followed by an interesting series of chapters (pp. 320–350) on the physical constitution of the Calabrias, the times and disturbed areas of the earthquakes, the sea-waves and after-shocks, the effects of the earthquakes on human beings, the epidemics following the earthquakes, and the number of victims. Sarconi notices that the earthquake of 28 Mar. disturbed an area greater than that affected by any of the earlier shocks, and he also points out the total want of order (that is, of periodicity) in the after-shocks or repliche. Some of the illustrations that accompany his report have become widely known by their reproduction in Lyell's chapter on the Calabrian earthquakes†.

31. Sir William Hamilton (1730–1803), not the least honoured bearer of a great name, was in 1764 appointed as British envoy extraordinary and plenipotentiary at the court of Naples. Much of his leisure time was spent in the study of the volcanic phenomena of the district, a study that led to his great work on the "Campi Phlegraei" (1776) and to a valuable account of the eruptions of Vesuvius in 1776 and 1777. In Feb. 1783, he visited Calabria to observe the effects of the recent earthquakes. His memoir, according to Dolomieu, "is the perception of a good observer, who had but a few instants to spare for examination on his trip to Calabria." In 1791, he married Amy Lyon, better known under the name of Emma Hart. He returned to England in 1800, the last years of his life being darkened by her relations with Lord Nelson‡.

* *Biografia degli Italiani Illustri*, vol. 1, 1834, pp. 263–266.

† *Principles of Geology*, 1st ed. 1830, Figs. 19, 20, 22–26.

‡ *Dict. Nat. Biog.* vol. 8, 1886, pp. 224–227.

Interesting as Hamilton's memoir is*, it cannot be said to add much to our knowledge of earthquakes, though his observations amply confirm those of other investigators. In his list of the disastrous shocks, he omits the two earthquakes of 7 Feb. and includes one on 27 Feb., but he notices that the earthquakes of 5 Feb. and 28 Mar. far exceeded the others in violence and were the only shocks that were sensible in Naples. He also records the fact, patent to all observers, that these great shocks affected different parts of the province. In other words, he realised the migration of the focus. He describes the translation of large masses of earth. "Sometimes," he says, "I met with a detached piece of the surface of the plain (of many acres in extent) with the large oaks and olive-trees, with lupins or corn under them, growing as well, and in as good order at the bottom of a ravine, as their companions, from whom they were separated, do on their native soil on the plain, at least 500 feet higher" (pp. 189–190). He points out the occurrence of series of parallel fissures in those parts of plains that adjoin ravines, and ascribes them correctly to "the earth rocking with violence from side to side, and having a support on one side only" (pp. 187–188, 198). Lastly, like other investigators, he notices how the amount of damage depended on the nature of the site, how buildings situated on high ground composed of a gritty sandstone suffered much less than those founded on the plain, which consists of a sandy clay. Indeed, houses built on the plain were universally levelled with the ground (pp. 179, 206).

32. The last of the original investigators of the Calabrian earthquakes, and the one whose account to English readers is perhaps the best known, was Déodat de Dolomieu (1750–1801)†. When only eighteen years of age, he killed a fellow-officer in a duel, was condemned to death, but was reprieved on account of his youth. The nine months that he spent in prison were devoted to the study of the physical sciences, and, on his release, he resigned his commission, travelled in Portugal, Spain, Sicily and the Pyrenees, and, in 1784, went to Italy in order to study the

* *Phil. Trans.* vol. 73, 1783, pp. 169–208.

† Dolomieu's full name was Déodat Guy Sylvain Tancrède de Gratet de Dolomieu.

Calabrian earthquakes. A series of contrary winds detained him during the whole of the months of February and March, obliging him to touch at nearly all the towns on the western coast of the province. From these he made excursions into the interior and examined all the ruined places. Of his *Mémoire sur les tremblemens de terre de la Calabre pendant l'année* 1783 (Rome, 1784), comparatively few copies were printed, and our knowledge of it is mainly due to its translation and inclusion in Pinkerton's *Voyages and Travels**. In 1791, he described the mineral *dolomite*, named after him. His later life was unfortunate, and his scientific career almost ended, as it had begun, in prison. In 1798, he joined Napoleon's expedition to Egypt, and during his return homewards owing to ill-health in 1799, he was driven by bad weather into Taranto and was taken prisoner. After nearly two years' confinement in a Neapolitan dungeon, he was released on the conclusion of peace in March 1801. He was then appointed professor of mineralogy at the Museum of Natural History in Paris, but his health never recovered from the hardships of his imprisonment, and he died about six months after his return to France†.

Dolomieu's account of the Calabrian earthquakes may be regarded as supplementing the reports of Grimaldi, Vivenzio, Sarconi and Hamilton. It is neither a chronicle of the earthquakes nor a record of disaster, but rather a study of the geological relations of the earthquakes. (i) Dolomieu notices, for instance, how small was the area affected considering the violence of the shocks. "Its limits," he says, "were the extremity of Calabria Citra on the one side; eastward it exercised no great ravage beyond Cape Colonne; nor westward beyond the town of Amanthea. Messina is the only town in Sicily which participated the disasters of the continent....The space, therefore, on which this terrible scourge displayed itself, was a length of thirty

* *A dissertation on the earthquakes in Calabria Ultra in the year* 1783, by the Commander Deodatus de Dolomieu, vol. 5, 1809, pp. 273–297. The memoir was also translated into Italian (Rome, 1784) and German (Leipzig, 1789).

† *Biographie Moderne*, vol. 1, 1811, pp. 378–379; *Bibl. Univ. Biog.* vol. 14, 1855, cols. 470–473; *Dictionnaire Général de Biographie et d'Histoire*, pt. I, p. 870.

leagues by the whole breadth of Calabria." (ii) Small though the disturbed area was, the variations of intensity within it were extremely rapid. Polistena, for example, was levelled with the ground on 5 Feb. On the same day, only four miles away, S. Giorgio suffered little, though later, on 7 Feb. and 28 Mar., it was seriously damaged. (iii) Like other investigators, Dolomieu was struck with the dependence of the intensity of the shock on the nature of the site. On the plain, which is intersected by deep ravines, near which the towns were usually established, the buildings were totally destroyed, while houses in the neighbourhood, built on the hard granite of the hills, escaped with slight injury. (iv) Dolomieu not only realised the displacement of the foci in successive earthquakes, but traced the direction of their migrations. The focus* of the first great earthquake, that of 5 Feb., was, he says, beneath the Calabrian plain in the neighbourhood of Terranova, Oppido and S. Cristina. That of the second, during the night of 5–6 Feb., was near Scilla. By the earthquakes of 7 Feb., other places were destroyed. "It seemed as if the focus or centre of explosion had ascended six or seven leagues higher up towards the north, and placed itself beneath the territory of Soriano and Pizzoni."† In the great earthquake of 28 Mar., which Dolomieu, like Grimaldi and Sarconi, regarded as the most violent of the series, "the centre of explosion changed for a third time, and again ascended seven or eight leagues higher towards the north," the seat of greatest violence being near Girifalco.

THE CHILIAN EARTHQUAKES OF 1822 AND 1835

33. In the works of several ancient writers, Seneca, Pliny, etc., we find accounts of changes of elevation referred to earthquakes. Some of the changes recorded may have occurred

* So far as I know, the first to use the word *focus* was É. Bertrand in 1757 (arts. 8–11). Dolomieu writes of the *focus* or *centre of explosion*, but he is careful to note that his use of either term is "not to indicate the cause, but only to explain the effect."

† "It is highly worthy of remark," says Dolomieu, "that the earthquake of the 7th of February was felt the most at Messina and Soriano, places very distant from each other; whilst it was mostly less violent in all the intermediate country." The explanation of this peculiarity is that there were two great earthquakes on 7 Feb., one at 8.20 p.m., near Monteleone and Soriano, the other, at about 10 p.m., near Messina.

during earthquakes, but in many, if not in most, cases, no clear distinction seems to have been drawn between seismic and volcanic phenomena, and this even as late as the close of the seventeenth century*. With the Chilian earthquakes of 1822 and 1835, however, we find less doubtful evidences of the rise of land described by several witnesses of skill and experience.

At the time of the Valparaiso earthquake of 19 Nov. 1822, Mrs Maria Graham (1785–1842) was living at Quintero, about thirty miles north of Valparaiso. An accomplished narrator of her various travels, she afterwards, as Lady Callcott, became widely known to a past generation as the author of *Little Arthur's History of England* (1835)†.

Early in 1824, Mrs Graham described the earthquake in a letter communicated to the Geological Society. It is an interesting account and has been often quoted. Further details, including especially the chronicle of the after-shocks, are given in her Chilian journal‡.

34. That Mrs Graham was a close and careful observer is clear from the details that she records. She noticed, for instance, the swelling of the rivers and the rise of the water-level in lakes, which she attributed to the vast masses of snow dislodged from the mountains. At the roots of all the trees, she could put her hand into the hollows formed round them by the rocking of the trunks. The promontory of Quintero consists of granite covered by sandy soil, the granite on the beach being intersected by parallel veins. After the great earthquake, the whole rock was found rent by clefts "sharp and new, and easily to be distinguished from the older ones." Many of these clefts could be traced for a mile and a half across the promontory. Lastly, houses built on granite stood tolerably well, while everything erected on sand or clay was damaged.

The accuracy of such details creates confidence in Mrs Graham's important observations on the rise of the adjoining coast.

* Robert Hook, *Posthumous Works*, ed. by R. Waller, 1705, pp. 299–302.

† *Dict. Nat. Biog.* vol. 8, 1886, p. 258.

‡ *Geol. Soc. Trans.* vol. 1, 1824, pp. 413–415; *Journal of a Residence in Chile during the year* 1822, 1824, pp. 305–339.

"It appeared on the morning of the 20th," she says, "that the whole line of coast from north to south, to the distance of above 100 miles, had been raised above its former level. I perceived from a small hill near Quintero, that an old wreck of a ship [the *Aquila*] which before could not be approached, was now accessible from the land, although its place on the shore had not been shifted. The alteration of level at Valparaiso was about 3 feet, and some rocks were thus newly exposed, on which the fishermen collected the scallop shell-fish, which was not known to exist there before the earthquake. At Quintero the elevation was about 4 feet. When I went [nearly three weeks later] to examine the coast..., although it was high water, I found the ancient bed of the sea laid bare and dry, with beds of oysters, muscles, and other shells adhering to the rocks on which they grew, the fish being all dead, and exhaling most offensive effluvia. I found," she adds, "good reason to believe that the coast had been raised by earthquakes at former periods in a similar manner; several ancient lines of beach, consisting of shingle mixed with shells, extending in a parallel direction to the shore, to the height of 50 feet above the sea."*

35. Mrs Graham's account is supported by John Miers (1789–1879), and rather warmly contested by Hugh Cuming (1791–1865). Both naturalists were at the time living at or near Valparaiso. According to Miers, "all the line of coast to the extent of fifty miles was raised nearly three feet above its former level; in some places the rocks on the shore were raised four feet. All around Quintero,...the fishermen had employed themselves digging shells for lime-making from a stratum four or five feet thick...at the height of fifteen feet above the usual level of the sea....This stratum was now again raised at least three feet."†

36. On the other hand, Cuming avers that neither he nor his friends noticed any change of level. (i) On the morning of 20 Nov., he observed the effects of the sea-waves on the Valparaiso beach, "but found nothing more than a high tide." (ii) In his pursuit of conchology, he was well acquainted with the rocks

* Readers of *Robinson Crusoe* (written a century earlier) will remember that, on completing the defences of his cave in the Island of Despair, the hero was driven into the open country by a series of violent earthquakes. The forecastle of his old ship, he found, "was heaved up at least six foot;... and the sand was thrown up so high on that side next her stern, that whereas there was a great place of water before, so that I could not come within a quarter of a mile of the wreck without swimming, I could now walk quite up to her when the tide was out" (D. Defoe, *The Life and Strange Surprising Adventures of Robinson Crusoe*, Heinemann, 1900, pp. 95–96). The first volume of the story was published in April 1719, and the above details were probably suggested by the eruption in St Vincent during the preceding summer.

† *Travels in Chile and La Plata*, vol. 1, 1826, pp. 394–395.

along the shore, but he never saw the least difference in their appearance, nor did he find any fuci, chitons, balani, etc., on the rocks, except in places covered by the tide. (iii) Vessels occupied the same anchorage after and before the earthquake, and "nautical men affirmed that there was not the least difference in the depth of the water in any part of the bay."*

37. After the lapse of more than a century, it is not easy to reconcile evidence so conflicting. It seems, however, to be of unequal value. (i) The three accounts were written at different times. The earthquake occurred on 19 Nov. 1822. Mrs Graham's letter was dated 4 Mar. 1824 and was founded on her journal for 20 Nov. 1822 and following days. Miers' *Travels* were written in 1825 and published in 1826. Cuming's letter was read to the Geological Society thirteen years after the earthquake, on 2 Dec. 1835; and he did not know of the alleged rise of land until after the publication of Mrs Graham's *Journal* in 1824, or about two years after the earthquake. (ii) Mrs Graham's minute observations have already been noticed. Cuming's evidence, in part at any rate, is less convincing. Sea-waves that, on their withdrawal, left small vessels stranded that before were afloat even at "dead low water," must have swept in beyond high-water mark. A difference of half a fathom in the depth of the water, unless actually measured, might pass unnoticed by seamen. The strongest point in Cuming's evidence is the observed absence of shell-fish above the level of the water at Valparaiso. He makes, however, the same observation with regard to Quintero, where Mrs Graham, who was accompanied by Lord Cochrane, found dead shell-fish above the level of high-water, still adhering to the rocks on which they grew and "exhaling most offensive effluvia."

Thus, the evidence for the Quintero district seems to favour Mrs Graham's account; and her observations clearly imply an actual elevation of the solid crust, though the extent of coast so affected, so far as it does not rest on her own authority, may remain in doubt†.

* *Geol. Soc. Trans.* vol. 5, 1840, pp. 263–265.

† See her *Journal*, p. 310, as regards the observations at Valparaiso.

38. This conclusion does not, however, involve the rejection of Cuming's observations at Valparaiso. The epicentral area cannot have been far from the coast, for the sea-waves swept in before the first shock ended or within three minutes of the beginning. Variations in the amount of the vertical displacement might thus be perceptible on land, and it is possible that the places examined by Cuming may have lain in areas of small or zero uplift*.

39. The next great earthquake in Chili occurred rather more than twelve years later, on 20 Feb. 1835, when Robert FitzRoy (1805–65) and Charles Darwin (1809–82) were in the neighbourhood during the voyage of H.M.S. *Beagle*†. At Concepcion, a rocky shoal, formerly submerged and afterwards exposed, showed that the uplift during the earthquake was two or three feet. The elevation, or the greater part of it, was, however, merely temporary, for, by the following July, the rock was covered at ordinary high-water, as in 1834. The chief upheaval took place in the island of Santa Maria, about 30 miles south-west of Concepcion, where, according to FitzRoy who visited the island twice, the rise was 8 feet at the south end, 9 feet in the middle, and 10 feet at the north end, beds of dead mussels being found at these distances above high-water mark. An extensive rocky flat lies round the north end of the island. Before the earthquake, it was covered by the sea, some projecting rocks alone being seen. In April, the whole flat was exposed, large portions "were covered with dead shell-fish, and the stench arising from them was abominable." By this time, the first elevation, as at Concepcion, had been reduced, for, according to

* During the Assam earthquake of 1897, vertical displacements occurred along the Chedrang fault. Within a distance of about 12 miles, there were three undulations (separated by intervals of no displacement) in which the maximum uplifts were 25, 35, and 32 feet, respectively (R. D. Oldham, *India Geol. Surv. Mem.* vol. 29, 1899, pp. 138–147). In the Alaskan earthquakes of 1899, the uplift along the west coast of Disenchantment Bay decreased from 42 ft. to 9 ft. 4 in. within little more than a mile (R. S. Tarr and L. Martin, *U.S. Geol. Surv. Prof. Paper* no. 69, 1912, p. 31 and plate 14; art. 160).

† R. FitzRoy, *Geog. Soc. Jl.* vol. 6, 1836, pp. 319–331, and *Narrative of the Surveying Voyages of H.M.S. Adventure and Beagle*, vol. 2, 1839, pp. 412–414, 420; C. Darwin, *Geol. Soc. Trans.* vol. 5, 1840, p. 619, and *Naturalist's Voyage*, etc., 1879, p. 310; A. Caldcleugh, *Phil. Trans.* 1836, pp. 24–25.

the inhabitants, their little port for three or four weeks after the earthquake was much shallower than it was in April. During the three days following the shock, the earth at Concepcion was never quiet for long and, from 20 Feb. to 4 Mar., more than 300 shocks were counted. It is probable that these after-shocks were caused by the gradual settling back of the crust after the first great uplift*.

FORERUNNERS OF PERREY AND MALLET

40. During the twelve years that preceded the coming of Perrey and Mallet, several geologists made useful additions to our knowledge of earthquakes.

First to be mentioned among these forerunners is Peter Merian (1795–1883), professor of physics and chemistry in the university of Basel and author of many papers on geology spread over a period of more than sixty years. Among them are several brief notes on earthquakes; but more valuable than any is a pamphlet published in 1834 on the earthquakes felt at Basel from 1020 to 1830†. During these eight centuries, there were 122 earthquakes of which the year is known, and, of all but four, the month. The pamphlet is a remarkable one, if only for the reason that it contains the first of a long series of notices on the winter preponderance of earthquakes. Merian gives the following monthly numbers for the 118 earthquakes:

Jan.	Feb.	Mar.	Apr.	May	June	July	Aug.	Sep.	Oct.	Nov.	Dec.
12	14	6	5	11	3	7	8	12	11	14	15

or, in winter (Dec.–Feb.) 41, spring (Mar.–May) 22, summer (June–Aug.) 18, and autumn (Sep.–Nov.) 37; or, again, in autumn and winter 78, in spring and summer 40.

He then proceeded to test the generality of this law by reference

* With the great Japanese earthquake of 1 Sep. 1923, most of the west coast of Sagami Bay was raised by amounts that ranged as high as 24 ft. 11 in. Soon afterwards, the land began to sink, and, two or three weeks later, the uplift on Hatsushima (a small island to the south-east of Atami) was reduced from 13 ft. 1 in. to 7 ft. 3 in., at Manazaru (in the north-west of the bay) from 12 ft. 10 in. to 8 ft. 6 in., at Misaki (at the southern end of the Miura peninsula) from 24 ft. 11 in. to 4 ft. 7 in. and, near the southern end of the Boso peninsula, at Tateyama from 8 ft. 10 in. to 4 ft. 7 in., at Banda from 14 ft. 9 in. to 5 ft. 3 in., and at Shirahama from 14 ft. 9 in. to 5 ft. 11 in. (*Geogr. Jl.* vol. 65, 1925, pp. 50–51).

† *Ueber die in Basel wahrgenommenen Erdbeben*, Basel, 1834, 20 pp.

to two other catalogues. From Hoff's annual lists for the years 1821–29 (art. 41), he found the following numbers: winter 36, spring 14, summer 19, and autumn 29; or autumn and winter 65, spring and summer 33. The list of earthquakes in northern Europe from 1775 to 1806 compiled by Louis Cotte (1740–1815)* suggests a less clearly marked grouping, namely, winter 44, spring 26, summer 40, and autumn 34; or, autumn and winter 78, spring and summer 66. On the whole, then, with the materials before him, Merian was clearly justified in upholding an increase in frequency during the winter months†.

41. The foundation for much of the work afterwards done by Perrey and Mallet was laid by Karl Ernst Adolf von Hoff (1771–1837), during a busy official life. He was the first to issue annual lists of earthquakes, the first also to compile a general catalogue of earthquakes for the whole world, at any rate on a considerable scale. His annual lists, as originally published, are ten in number and relate to the years 1821–30‡. Those for the years 1831 and 1832 appeared later in his *Chronik der Erdbeben und Vulkan-Ausbrüche*§. Taking all the twelve years together, the annual numbers for the whole world range from 17 to 95, the total number being 555.

The winter preponderance noticed by Merian is confirmed by the figures given by Hoff for the earthquakes of the Northern Hemisphere during the ten years 1821–30, namely‖:

Jan.	Feb.	Mar.	Apr.	May	June	July	Aug.	Sep.	Oct.	Nov.	Dec.
31	36	31	29	33	33	20	31	24	41	26	34
Win. 98			Spr. 95			Sum. 75			Aut. 101		

Hoff's most important work on earthquakes was his *Chronik der Erdbeben.* If less complete than Mallet's catalogue (art. 67),

* *Jl. Phys.* vol. 65, 1807, pp. 161–168, 329–364.

† Merian returned to the subject a few years later (Basel, *Bericht*, vol. 3, 1838, pp. 65–78), with slightly increased numbers but the same results.

‡ *Ann. Phys. Chem.* vol. 7, 1826, pp. 159–170, 289–304; vol. 9, 1827, pp. 589–600; vol. 12, 1828, pp. 555–584; vol. 15, 1829, pp. 363–383; vol. 18, 1830, pp. 38–56; vol. 21, 1831, pp. 202–218; vol. 25, 1832, pp. 59–91; vol. 29, 1833, pp. 415–447; vol. 34, 1835, pp. 85–108.

§ Two vols. Gotha, 1840–41, 876 pp.

‖ *Ann. Phys. Chem.* vol. 34, 1835, p. 104.

it resembles it in many ways and is a very valuable and praiseworthy piece of work. It begins, like Mallet's, with the year 1606 B.C. and closes with the year 1805. For the next fifteen years, there are no entries, with the exception of a few in 1820. Then follow the annual lists for the years 1821–32. The total number of earthquakes chronicled by Hoff amounts to 2225.

42. P. N. C. Egen (1793–1849), for many years a teacher of mathematics and physics at Elberfeld and elsewhere, wrote one memoir on earthquakes, in which he described the Netherlands earthquake of 23 Feb. 1828*. It is of interest as it contains the first attempt to devise an arbitrary scale of seismic intensity, one that closely resembles the lower degrees of the Rossi-Forel scale (art. 101), which it forestalls by half a century. The scale is as follows:

EGEN SCALE OF INTENSITY

1. Only very slight traces of the earthquake are sensible.
2. A few persons, under favourable conditions, feel the shock; glasses close together jingle, small plants in pots vibrate; hanging bells are not rung.
3. Windows rattle, house-bells are rung; most persons feel the shock.
4. Slight movement of furniture; the shock in general so strong that it is felt by everyone.
5. Furniture shaken strongly, walls are cracked, only a few chimneys thrown down, the damage caused being insignificant.
6. Furniture shaken strongly; mirrors, glass and china vessels broken; chimneys thrown down, walls cracked or overthrown.

On the map that accompanies his paper, Egen underlined the places at which the intensity was 6 or 5 in red, those at which it was 4 or 3 in blue, and those at which it was 2 or 1 in yellow. Isoseismal lines are not shown on the map; indeed, the observations are too few to allow them to be drawn accurately, but it seems probable that Egen drew at least the innermost isoseismal, for he remarks that the places at which the intensity was 6 lie

* *Ann. Phys. Chem.* vol. 13, 1828, pp. 153–163.

within an elliptical area, the longer axis of which is directed east and west. So far as I know, Egen's map is the first in which any attempt is made to depict the variation in the intensity of an earthquake throughout its disturbed area.

43. The year 1830 marks an epoch in the history of geology, for it was in this year that Charles Lyell (1797–1875) published the first volume of his *Principles of Geology*. Frequently enlarged in subsequent editions (the twelfth and last appearing in 1875), the structure of the four chapters on earthquakes* remained fundamentally the same throughout. In them, for the first time, we have a series of great earthquakes described from a scientific point of view. Beginning with the Murcian earthquake of 1829 and ending with the Jamaica earthquake of 1692, he refers to 35 earthquakes, mostly of destructive strength. To the Calabrian earthquakes of 1783, a complete chapter (pp. 475–501) is devoted. They are the only earthquakes "of which the geologist can be said to have such a circumstantial account as to enable him fully to appreciate the changes which this cause is capable of producing in the lapse of ages." If later studies of the Calabrian earthquakes† have added further details, they do not lessen the value of Lyell's remarkable chapter.

In later editions of the *Principles*, brief accounts of some earthquakes are omitted, and of others are inserted, the principal additions being those of the Chilian earthquake of 1835 and the New Zealand earthquake of 1855‡. In both of these, we have

* Vol. 1, 1830, pp. 457–553.

† G. Mercalli, Roma, *Soc. Ital. Mem.* vol. 11, 1897, pp. 30–45; M. Baratta, *I Terremoti d' Italia*, pp. 268–292, 813–823.

‡ Fifth ed. 1837, vol. 2, pp. 183–188; tenth ed. 1867, vol. 2, pp. 82–89. It may be of interest to notice Lyell's treatment of the elevation of the Chilian coast in 1822 (arts. 33–38). Mrs Graham's account is quoted in the first and all later editions. The supporting evidence of Cruickshanks is given in the second edition (1832), that of Freyer and Meyen in the fourth (1835). In the latter, he also quotes Cuming's adverse evidence (given personally), though he considers the fact of elevation established by "the testimony of many witnesses," and adds "although the change of level may perhaps be found to have been less uniform in different places than some have assumed" (vol. 2, p. 234). In the fifth edition (1837), he referred to Cuming's letter read before the Geological Society, but "felt satisfied with the proofs of elevation" (vol. 2, p. 192). In the sixth edition (1840), Cuming's objections are quoted, but much less fully. In the seventh and all later editions (1847–75), all reference to them is omitted.

evidence, if such were needed, of Lyell's unfailing interest in the changes of elevation that accompany great earthquakes*.

44. David Milne, afterwards David Milne Home (1805–90), the son of Admiral Sir David Milne, G.C.B., belonged to that small band—those who are relieved by ample fortune from the cares of this world—to whom science in this country is so deeply indebted. At an early age, he was interested in geology, and, while studying law at Edinburgh university, he attended the lectures of the professor of natural history, Robert Jameson (1774–1854). In 1832, he married Jean Forman Home and afterwards assumed her name. Beginning work as an advocate in 1826, he abandoned the practice of law in 1845, and thenceforth devoted himself to the management of his estates, to county business, and, above all, to scientific pursuits. His more important memoirs deal with the geology of Roxburghshire and the Mid-Lothian and East-Lothian coalfields, the origin of the parallel roads of Glen Roy†, the distribution of boulders in Scotland, and the study of earthquakes in Great Britain. "He was one of the true aristocrats—not merely a possessor of position, wealth, and lands, but of knowledge, public spirit, ability, and intellect."‡

Milne's interest in earthquakes was roused by the remarkable shocks which occurred at and near Comrie in 1839. In 1840, probably at his suggestion, the British Association appointed a committee for the study of earthquakes in this country. Milne acted as secretary to the committee, and wrote the four reports which were presented in the years 1841–44§. One of the members

* In addition to the chapters on earthquakes in the *Principles*, Lyell wrote an interesting account of his visit in March 1846 to the central area of the New Madrid earthquakes of 1811–12 (*Second Visit to the United States*, vol. 2, 1849, pp. 229–239). Many effects of the earthquake were then, as they still are, distinctly visible, such as "sand-blows," fissures in the soil, and the "sunk country," with its many dead trees supposed to have been killed by the loosening of the roots during the frequent passage of the earthquake-waves (art. 159).

† See *Life and Letters of Charles Darwin*, vol. 1, 1887, pp. 361–363; *More Letters of Charles Darwin*, vol. 2, 1903, pp. 172–188.

‡ *Biographical Sketch of David Milne Home*, by his daughter G. M. H. 1891; *Edin. Geol. Soc. Trans.* vol. 6, 1893, pp. 119–127; *Geol. Soc. Quart. Jl.* vol. 47, 1891, *Proc.* p. 59.

§ *Brit. Ass. Rep.* 1841, pp. 46–50; 1842, pp. 92–98; 1843, pp. 125–127; 1844, pp. 85–90.

of the committee was James David Forbes (1809–68), the ideal student of science, already well known for his researches on heat, and soon to become famous as an investigator of glacial phenomena*. The work of the committee was confined almost entirely to the Comrie earthquakes—as it happened there were few others elsewhere to record—and, on their becoming slight and infrequent, the committee lapsed after the meeting of 1844. A few points of historical interest may be referred to here: (i) Milne's invention of the word *seismometer*, our oldest seismological term; (ii) Forbes' design of a seismometer, of which an inverted pendulum was the essential part†; and (iii) Milne's determination of the epicentre by two or more lines of direction.

During the years in which this committee was at work, Milne published a valuable series of *Notices of earthquake-shocks in Great Britain, and especially in Scotland*, copies of which with various additions were bound together and re-issued in 1887‡. The first part contains a register of 238 earthquakes felt in Great Britain from 1608 to 1839. This was founded mainly on papers published in various journals, but to a great extent also on the diary kept by Samuel Gilfillan (1762–1826), Secession Minister at Comrie, who noted the occurrence of at least 68 earthquakes from 1792 to 1822. The second and longer part deals with the Comrie earthquakes from 3 Oct. 1839 to the end of 1844, and includes a full account of the principal shock of 23 Oct. 1839—the most detailed report up to that time published on any British earthquake.

While this remarkable series of papers may be said to have founded the study of British earthquakes, it also contains one or two points of general interest. In the second paper (published in 1841), Milne gives a summary of the various phenomena of the earthquakes in his catalogue, laying perhaps undue stress on the weather-conditions at the times of their occurrence. He describes very clearly the transmission of the vibrations from

* *Life and Letters of James David Forbes*, by J. C. Shairp, P. G. Tait, and A. Adams-Reilly, 1873.

† *Edin. Roy. Soc. Trans.* vol. 15, 1844, pp. 219–228.

‡ *Edin. New Phil. Jl.* vol. 31, 1841, pp. 92–122, 259–309; vol. 32, 1842, pp. 106–127, 362–378; vol. 33, 1842, pp. 372–388; vol. 34, 1843, pp. 85–107; vol. 35, 1843, pp. 137–160; vol. 36, 1844, pp. 72–86, 362–377.

the origin outwards in spherical waves, and shows that at the epicentre the movement would be vertical, becoming more nearly horizontal with increasing distance from the epicentre.

"If instruments," he remarks, "could be invented which at different places would indicate, not merely the relative intensity of the shocks, but the direction in which they acted on bodies, means would be obtained of determining the point in the earth's interior from which the shocks originated"—a method of determining the depth of the focus that was afterwards devised independently by John Milne and used by him, by Omori and others to determine the depth of some Japanese earthquake-foci*.

Again, he notices that the places in Scotland that are most subject to earthquakes are characterised by certain geological features. He recalls how numerous are the earthquakes at Comrie and in the Great Glen, and adds "now, along these districts, it is well known that there are deep and extensive fissures and dislocations in the earth's crust."

45. This chapter would be incomplete without some reference to the work of Friedrich Heinrich Alexander, Baron von Humboldt (1769–1859), whose long life almost spanned the period covered by it. From 1799 to 1804, he travelled in Venezuela, Columbia, Ecuador, Peru and Mexico. On returning to Europe, twenty years (1807–27) were spent in Paris in drawing up the reports on this journey. From 1827, except for nine months in Central Asia, he lived in Berlin, and there wrote what some regard as his chief title to fame, the four volumes of *Cosmos* published in 1845–58†.

In the *Personal Narrative*‡ of his travels in South and Central America (1811–29), Humboldt describes two great earthquakes—the Cumana earthquake of 14 Dec. 1797 and the Caraccas earthquake of 26 Mar. 1812. The former occurred before his visit, though he felt after-shocks on 4 Nov. 1799, and the latter some years after he had returned to Europe. There are points of interest still in the account of the Cumana earthquake.

* *Edin. New Phil. Jl.* vol. 31, 1841, pp. 276–277; *Japan Seis. Soc. Trans.* vol. 1, pt. 2, 1880, pp. 104–105; *Tokyo, Coll. Sci. Jl.* vol. 11, 1899, pp. 194–195; *Japan Earthq. Inv. Com. Notes*, No. 6, 1924, p. 6; Kobe (Japan), *Marine Obs. Mem.* vol. 1, 1924, p. 145.

† K. C. Bruhns, *Alexander von Humboldt*, 3 vols. Leipzig, 1872; translated by J. and C. Lassell, 2 vols. London, 1873.

‡ (Translated by H. M. Williams), vol. 2, 1814, pp. 214–238; vol. 3, 1818, pp. 316–327; vol. 4, 1819, pp. 1–55.

Humboldt refers to the prevalent belief in a connexion between earthquakes and the state of the weather and urges that it has no real foundation. The rapidity with which the undulations travel is proof, in his opinion, that the "centre of action" is remote from the surface. He realises the migration of earthquake-foci and attributes it to the opening of new underground communications. He believes in the intimate connexion between earthquakes and volcanic eruptions. "Everything in earthquakes seems to indicate the action of elastic fluids seeking an outlet to diffuse themselves in the atmosphere," and thus he is led to the view so widely held that volcanoes act as safety-valves for the surrounding country (arts. 17, 166).

46. At the age of 75, "in the late evening of an active life," Humboldt published the first volume of *Cosmos*—"the great work of our age," according to a contemporary—in which he sought to portray "the unity amid the complexity of nature." It is in this volume and the last that the sections on earthquakes are included. Humboldt's own experience, gained more than forty years before, is to some extent the foundation of the first chapter. Notable as one of the early attempts to give an account of earthquakes in general, it contains two features that are still of interest. He anticipated Mallet in his views on the propagation of earthquake-motion, which, he says, is "most generally effected by undulations in a linear direction, with a velocity of from 20 to 28 miles a minute, but partly in circles of commotion or large ellipses, in which the vibrations are propagated with decreasing intensity from a centre towards the circumference." The circles of commotion from two distinct centres may coalesce. "We may even suppose *interference* to exist here as in the intersecting waves of sound." Of great interest also is Humboldt's account of the earth-sounds or *bramidos* of Guanaxuato in Mexico, which began on 9 Jan. 1784.

In the last volume, the section on earthquakes comes under the heading "Reaction of the interior of the earth upon its surface," and, as it was published in 1858, he was able to refer to the work of Mallet, Hopkins, and others. Perhaps the most noteworthy point in this section is Humboldt's clear distinction between *plutonic* and *volcanic* earthquakes. "The most destructive earthquakes recorded in history," he remarks, "stand in no connexion with the activity of volcanoes....By far the greater part of the earthquakes upon our planet must be called Plutonic."*

CONCLUSION

47. As already remarked (art. 25), the period of eighty years (1760–1840) embraced in this chapter produced no outstanding figure in seismology; yet it was one of steady growth in our knowledge of earthquake-phenomena. To realise the advance made, we have only to recount the principal achievements of the numerous occasional workers. We have, for instance, (i) the first series of annual lists of earthquakes (art. 41), (ii) the first notable catalogue of earthquakes for the whole world (art. 41), and (iii) the discovery of the preponderance of earthquakes in the winter months (art. 40). Again, we can point to (iv) the first scientific investigation of a series of great earthquakes (arts. 26–32), (v) the first attempt to trace the variation of intensity throughout the disturbed area of an earthquake (art. 42), and (vi) the first detailed lists of the after-shocks of a great earthquake (art. 29). Not less important are (vii) the observations on the changes of elevation during earthquakes (arts. 33–39), (viii) the tracing of the migration of the epicentre during successive great earthquakes (art. 32), and (ix) the recognition of the dependence of the amount of damage on the nature of the underlying soil (arts. 32, 34). Lastly, (x) the period includes the first attempt to compile a bibliography of earthquake-literature (art. 3).

* *Cosmos* (translated by E. C. Otte and W. S. Dallas), vol. 1, 1849, pp. 199–213; vol. 5, 1858, pp. 165–183.

CHAPTER IV

ALEXIS PERREY

48. Few men have lived the life of a student more consistently and more earnestly than Alexis Perrey (1807–82). He was born on 6 July 1807, at Sexfontaines, a small village in the department of Haute-Marne, where his father, Nicolas Perrey, was a forest-ranger. In 1823, he was admitted to the college of Langres, to be trained for holy orders. Two years later, he received minor orders, but, finding himself unable to accept major orders, he closed his theological studies in 1829, and, after the revolution of 1830, definitely abandoned the clerical life. Maintaining himself as a tutor, first at Nancy and afterwards in Paris, he took the degree of bachelor of letters at Nancy in 1829, and bachelor and licentiate of sciences at Paris in 1831. In 1832, he began his career as lecturer in mathematics at the royal colleges of Angers, Amiens, Cahors, and (in 1837) of Dijon. At the latter place he lived for thirty years, becoming deputy professor of pure mathematics in 1838, director of the municipal observatory in 1844, and professor of applied mathematics (a chair created for him) in 1847.

Perrey's early papers (1829–41) were brief notes on meteorology, astronomy and pure mathematics, and it was not until 1841 that he touched on the subject that was to form the main work of his life. To this year belongs a short paper on the earthquakes from A.D. 306 to 1583 and on their variation in frequency throughout the year. Written as a letter to Arago (1786–1853), it was communicated by him to the Academy of Sciences. From this time onwards, his interest in earthquakes never failed, and his work, aided by the encouragement and active help of such men as Arago, Élie de Beaumont (1798–1874) and Quetelet (1796–1874)*, was pursued, but for one interruption through illness (1854–55), until the year 1875.

* How much Perrey owed to Quetelet, the perpetual secretary of the Royal Academy of Belgium, is evident from the long series of memoirs

Perrey's devotion to science was so intense that his health began to suffer from the strain. Towards the close of 1867, he resigned his professorship and other offices, left Dijon, and retired to Lorient in Brittany. Here, for several years, he continued to issue his annual lists of earthquakes, but the preparation of the regional memoirs had to be abandoned owing to the distance from libraries. In October 1875, Perrey presented his last memoir to the Academy of Sciences. In this, he considered his first and second laws of earthquake-frequency in connexion with his thirty annual lists of earthquakes from 1843 to 1872. He was, however, obliged to leave the memoir unfinished, the calculations for the third law being prevented by "the precarious state of his sight." In 1881, he removed to Paris, and died there after a brief illness on 29 Dec. 1882*.

Excluding rather more than a score of papers on miscellaneous subjects—chiefly meteorology and astronomy—Perrey's work falls readily into five divisions: (i) the annual periodicity of earthquakes, (ii) the annual lists of earthquakes from 1843 to 1872 inclusive, (iii) the regional monographs, 21 in number, (iv) the lunar periodicity of earthquakes, and (v) the bibliography of seismology.

49. Annual Periodicity of Earthquakes. The winter preponderance of earthquakes was first noticed by Merian in 1834 (art. 40) and confirmed by Hoff in 1835 (art. 41). Perrey's attention was soon attracted to it, and four of his early papers deal with the subject†. Until the third paper appeared (1842),

(38 in number) published in the *Bulletin* or *Mémoires Couronnées* of that society. The last annual list (that for 1872) was communicated to the Academy in August 1875, after Quetelet's death, and was never printed.

* Perrey's death, only seven years after the publication of his last memoir, seems to have passed unnoticed, either in scientific journals or in the transactions of the scientific academies with which he was connected. Forty-two years after his death, the omission has been supplied by the publication of an interesting paper in the *Mémoires* of the Dijon Academy of Sciences (1924, pp. 1–73). This paper is in two parts. The analysis and bibliography of Perrey's work are contributed by E. Rothé, director of the central seismological bureau of Strasbourg, while the biographical section, from which most of the above details are taken, is written by Perrey's grandson, H. Godron.

† Paris, *Ac. Sci. C. R.* vol. 12, 1841, pp. 1185–1187; vol. 13, 1841, pp. 899–903; vol. 15, 1842, pp. 643–646; vol. 17, 1843, pp. 608–625.

there is no evidence that he was acquainted with Merian's and Hoff's results, and it is therefore probable that it was discovered independently by himself. From 1841 to 1843, he made for this purpose four lists of earthquakes, containing respectively 262, 987, 1329 and 2022 records. Is it not possible that this effort to improve his earthquake catalogues led him onward to the main work of his life?

In each of the four papers, he notices the marked preponderance of earthquakes in the winter months, the percentages of the total number of earthquakes in the successive lists that occurred during the six winter months being 59, 58, 58 and 58. The monthly and seasonal numbers given in the first two lists are as follows:

Interval	Jan.	Feb.	Mar.	Apr.	May	June	July	Aug.	Sep.	Oct.	Nov.	Dec.
306–1583	21	12	10	14	12	13	9	14	15	17	10	28
	Win. 43			Spr. 39			Sum. 38			Aut. 55		
306–1800	86	60	66	56	46	60	47	43	58	53	61	92
	Win. 212			Spr. 162			Sum. 148			Aut. 206		

The subject possessed an unfailing interest for Perrey and will be considered further in connexion with the regional memoirs*. In the last paper of the group, he touches on some other subjects. The greater number of earthquakes recorded during the nineteenth century he ascribes—and no doubt correctly—to the growth of the newspaper press. At the same time, the varying numbers of earthquakes in successive years show that the earth's crust "is not always in an identical state of oscillation." The mean yearly number of earthquakes at this time he finds to be 32·7. He also notices a point that has attracted some attention in recent years, namely, that countries separated by vast distances have been disturbed simultaneously: for example, Holland and Spain, Lisbon and Saxony, Calabria and Maurienne, Savoy and Scotland.

50. Annual Lists of Earthquakes. Shortly before Perrey began his annual lists, but probably unknown to him, similar

* Perrey also refers to this subject in the lists of earthquakes for the years 1843, 1844–47 and 1854, and in nearly every regional memoir gives the monthly numbers of earthquakes.

lists had been issued for the years 1821–32 by Hoff (1770–1837; art. 41), in which the yearly numbers (for the whole world) range from 17 to 95.

Perrey's annual lists for the 29 years 1843–71 occupy 28 papers*, the lists for 1845 and 1846 being included in one paper, as well as those for 1866 and 1867. Each of the papers from 1845 to 1871 includes supplements for previous years, and, between the lists for 1868 and 1869, Perrey published a paper containing additional notices of earthquakes from 1843 to 1868. The total number of pages in these 28 memoirs is just over 2500. A quite considerable part is devoted to a chronicle of the volcanic eruptions of the particular years, but the number of earthquakes described or catalogued is more than 21,000. And of this large number, almost exactly one half appeared in the first lists for the several years, and the other half in later supplements†.

As might be expected, Perrey seems to have found some difficulty in securing a suitable medium for the publication of his annual lists. Valuable as they are, they nevertheless occupy considerable space, and they can hardly be described as attractive reading. Nearly half of them contain more than a hundred pages, two of them more than two hundred. The first list, that for 1843, appeared in the *Comptes Rendus* of the Paris Academy of Sciences. The lists for the years 1844 to 1853 were published in the

* With the annual lists may also be included the memoir on the earthquakes and volcanic eruptions of the Hawaiian Islands in 1868 (Lyon, *Soc. Agric. Ann.* vol. 2, 1869, pp. 95–157), in which he gives the list of shocks observed by C. G. Williamson at Christchurch (Kona district).

† Brux. *Ac. Bull.* vol. 12 (pt. 2), 1845, pp. 329–335; vol. 15, 1848, pp. 442–454; vol. 16, 1849, pp. 323–329; vol. 17, 1850, pp. 216–235; vol. 18, 1851, pp. 291–308; vol. 19 (pt. 1), 1852, pp. 353–396; *ibid.* (pt. 2) pp. 39–69; vol. 20 (pt. 2), 1853, pp. 39–69; vol. 21 (pt. 1), 1854, pp. 457–489; vol. 22 (pt. 1), 1855, pp. 526–572; vol. 23 (pt. 2), 1856, pp. 23–68; vol. 1, 1857, pp. 64–128; Brux. *Mém. Couronn.* (8vo ed.), vol. 8, 1859, no. 3, 79 pp.; vol. 10, 1860, no. 4, 114 pp.; vol. 12, 1861, no. 4, 68 pp.; vol. 13, 1862, no. 3, 77 pp.; vol. 14, 1862, no. 3, 74 pp.; vol. 16, 1864, no. 5, 112 pp.; *ibid.* no. 6, 179 pp.; vol. 17, 1865, no. 5, 213 pp.; vol. 18, 1866, no. 4, 98 pp.; vol. 19, 1867, no. 3, 125 pp.; vol. 21, 1870, no. 5, 223 pp.; vol. 22, 1872, no. 3, 116 pp.; *ibid.* no. 4, 116 pp.; vol. 23, 1873, no. 6, 70 pp.; vol. 24, 1875, no. 3, 146 pp.; *ibid.* no. 4, 145 pp.; Dijon, *Ac. Sci. Mém.* 1843–44 (pt. 1), pp. 334–342; 1845–46, pp. 393–479; 1847–48 (pt. 2), pp. 68–115; 1849 (pt. 2), pp. 1–40; 1850 (pt. 2), pp. 51–71; vol. 1 (pt. 2), 1851, pp. 1–36; vol. 2 (pt. 2), 1852–53, pp. 1–65, 79–128; vol. 3 (pt. 2), 1854, pp. 1–46; Paris, *Ac. Sci. C. R.* vol. 18, 1844, pp. 393–403.

Mémoires of the Dijon Academy of Sciences, those from 1847 to 1853 were also reprinted (the main parts of the papers, not as a rule the supplements) in the *Bulletin* of the Royal Academy of Belgium. During the next two years, 1854 and 1855, the complete memoirs were issued in the latter journal alone; after which, from 1856 to 1871, they were published by the same Society among their *Mémoires Couronnées* (8vo edition).

51. It is interesting to notice how Perrey's work gradually expanded. In the first list, that for 1843, the number of earthquakes recorded for the whole of Europe and the adjoining portions of Africa and Asia is only 47, though the number for this year afterwards rose in successive instalments to 225. For the following year, the corresponding numbers were 35 and 308. From 1845 onwards, his lists embraced the whole world, and the numbers amounted to 107 and 274 in 1845, and 124 and 374 in 1846. After 1850, the total number never fell below 500 in any one year, and on five occasions it exceeded one thousand. When he retired from Dijon to Lorient, Perrey feared that his work would be hindered by his absence from libraries, and he endeavoured to supply this defect by corresponding with the recorders of earthquakes in foreign lands. That his efforts were successful is evident from the fact that the average annual number of earthquakes for the four years before retirement was 636, while that for the four years after was 1270, or just double.

Perrey's annual lists are a vast storehouse of facts. For the most part, he was content to leave the discussion of the results to others. In three lists (those for 1843, 1847 and 1854), he notices the seasonal distribution of earthquakes; in the list for 1847, he investigates the lunar periodicity of earthquakes, and, in that for 1854, he refers once more to the subject. The descriptions, it must be admitted, are often tantalisingly brief. There is seldom any attempt to determine the position of the epicentre, none, I think, to discover the relations between successive earthquakes or between the earthquakes and the geological structure of the central district. The fullest account that he gives of any earthquake is perhaps that of the Visp valley earthquake of 25 July 1855 (art. 121). In this, which occupies 20 pages, he summarises the work of several investigators—Collomb, Dufour, Heusser, Morlot, Nöggerath, Rion and Volger. He describes the five zones of successively diminishing intensity mapped in Volger's memoir (art. 121). A few observations on the direction of the shock are given, but perhaps the most valuable part of the account is the list of after-shocks which, up to the end of the year, numbered 293.

Such detail, however, is unusual in Perrey's lists. The task which he set himself was, in fact, too large to admit of elaboration. What we have to be thankful for is that, at a time when few scientific men regarded earthquakes as deserving of close attention, there was fortunately one man who was willing to devote his leisure, year after year, for a whole generation, to the uninspiring task of amassing facts of the utmost value for future workers. It may be the charge of a hewer of wood, of a drawer of water, but no one who has followed in Perrey's footsteps and availed himself of his ample materials will cease to be grateful for the self-sacrificing labour of his well-spent life.

52. Regional Memoirs. If Perrey had ended his labours with his annual catalogues, he would have "deserved well of science." But he further increased our debt to him by his memoirs on the earthquakes of special regions. These are 21 in number, and the labour involved in their preparation may be inferred from the facts that they occupy altogether nearly 2000 pages and chronicle more than 9000 earthquakes. The publication of the series began in 1845 and Perrey's "tâche penible," as he calls it, ended in 1866*. In the first year, he issued three of these memoirs, and four in each of the years 1846 and 1847, after which they appeared at longer intervals†.

* Even while living in Dijon, Perrey's work was hampered by the absence of certain memoirs, such as Merian's on the earthquakes of Basel and D. Milne's on those of the British Isles. In Lorient, the difficulty would be prohibitive, and this no doubt set the limit to his regional memoirs.

† Brux. *Mém. Couronn.* (4to), vol. 18, 1844–45, 110 pp. (France, Belgium and Holland); vol. 19, 1845–46, 113 pp. (Rhine basin); vol. 22, 1846–47, 145 pp. (Italy); vol. 23, 1848–50, 73 pp. (Balkan peninsula); *ibid.* (8vo) vol. 7, 1858, 134 pp. (Peru, Colombia and Amazon basin); Dijon, *Ac. Sci. Mém.* 1845–46, pp. 299–323 (Algeria and N. Africa), 325–392 (Antilles); 1847–48 (pt. 2), pp. 1–67 (Atlantic basin); vol. 8 (pt. 2), 1861, pp. 85–194 (Philippines); vol. 13, 1866, pp. 121–251 (Aleutian Is., Alaska, etc.); Lyon, *Ac.*

The regional memoirs deal with a large part of the habitable globe. They cover all Europe, except Russia and the parts of Central Europe outside the basins of the Rhine and the Danube. Asia and Africa are less completely represented. Syria is included with the Balkan peninsula, but, with the exception of the northern portions of both continents, Perrey confines himself to the island regions of the Pacific. India, Turkestan, China and the Malay Archipelago are left untouched. North and Central America are treated in three memoirs, and, in two others, the important seismic regions of Chili, and of Peru, Colombia and the basin of the Amazon. If New Zealand lies outside his purview, it must be remembered that its seismic record begins with the year 1848.

The chief difficulty encountered by one who, like Perrey, deals with the earthquakes of foreign lands, is that of estimating the trustworthiness of his authorities. The earthquakes of some of his seismic regions—for instance, Japan, Italy and Great Britain —have since been treated in considerable detail. In his memoir on the earthquakes of the British Isles, Perrey's list extends from 1048 to 1848, and, for Great Britain alone, he gives entries of 511 shocks. Not all of these, however, were true earthquakes, and, for various reasons, it seems to me probable that at least 91 of them were spurious. Thus, of every five shocks recorded by Perrey, four were probably of seismic origin.

53. In plan, the different memoirs present but little variation. A brief introduction and list of authorities are followed by the catalogue of earthquakes and volcanic eruptions, and this again by a summary of which the principal elements are the seasonal distribution of earthquakes and observations on the direction of

Sci. Mém. vol. 12, 1862, pp. 281–390 (Japan); Lyon, *Soc. Agric. Ann.* vol. 8, 1845, pp. 265–346 (Rhone basin); vol. 9, 1846, pp. 333–414 (Danube basin); vol. 10, 1847, pp. 461–510 (Spain and Portugal); vol. 1, 1849, pp. 115–177 (British Isles); vol. 6, 1854, pp. 232–437 (Chili); vol. 8, 1864, pp. 209–374 (Kurile Is. and Kamtschatka); Vosges, *Soc. Émul. Ann.* vol. 6, 1847, 37 pp. (Mexico and Cent. America); vol. 7, 1850, 62 pp. (United States and Canada); *Voyages en Scandinavie de la Com. Sc. du Nord, Géog. phys.* vol. 1, pp. 409–469 (Scandinavia); *Ann. Magnét. et Météor. du Corps des Ingénieurs de Russie, an.* 1846, pp. 201–236 (N. Europe and N. Asia). Perrey's tables on the monthly distribution of earthquakes and on the direction of the shock are reprinted in Mallet's fourth report (*Brit. Ass. Rep.* 1858, pp. 2–31).

the shock. Sometimes he discusses briefly the apparent relations between the occurrence of earthquakes and changes in the weather, and, once, in the memoir on the Rhone basin, he touches on the sound-phenomena, noticing the capricious audibility of the sound and its limitation to the district around the origin, referring also to the different types of sound—namely, those of a cannon, of thunder, an avalanche or the fall of a heavy mass, a storm or a strong gust of wind, and, most often, that of a heavily laden carriage passing rapidly on the road or over arches.

Few of these regional memoirs—and those only for which the records were obviously incomplete—appeared without a discussion of the annual periodicity of earthquakes. At the outset, he was met by the difficulty of defining the unit earthquake, and he solved it by regarding as a single earthquake, not every recorded shock, but all the shocks that occurred in a given district during a fortnight or even a month provided that there were no interruption of a week or more. When an earthquake-series lasted for several consecutive months, as in Calabria in 1783, he counts one for each month. On the other hand, he regards shocks as distinct if they occurred almost simultaneously in very distant localities, such, for instance, as Calabria or Sicily and Piedmont or Lombardy.

54. Having defined the unit earthquake, Perrey gives for every region the number of earthquakes in each month from century to century. The total monthly numbers are then divided by the average of all twelve numbers, and, in this way, the results from different regions become comparable. As a rule, they are represented by curves, which he calls *seismic curves*. For instance, the monthly numbers for the region consisting of France, Belgium and Holland are given in the second line of the following table*, and the quotients denoting the relative monthly frequency in the third line. The seismic curve is given in Fig. 2.

Jan.	Feb.	Mar.	Apr.	May	June	July	Aug.	Sep.	Oct.	Nov.	Dec.
83	64	53	55	42	36	47	40	50	48	60	78
1·52	1·17	·97	1·01	·77	·66	·86	·73	·91	·88	1·09	1·43

* In obtaining these monthly numbers, Perrey omits certain more or less long-lived series of earthquakes.

These monthly numbers are again grouped in three ways (i) in meteorological seasons, (ii) in winter and summer half-years, and (iii) in pairs of months including the solstices and equinoxes. Thus, for the same district, we have

Winter Jan.–Mar.	Spring Apr.–June	Summer July–Sep.	Autumn Oct.–Dec.
200	133	137	186
1·22	·81	·84	1·13

also,

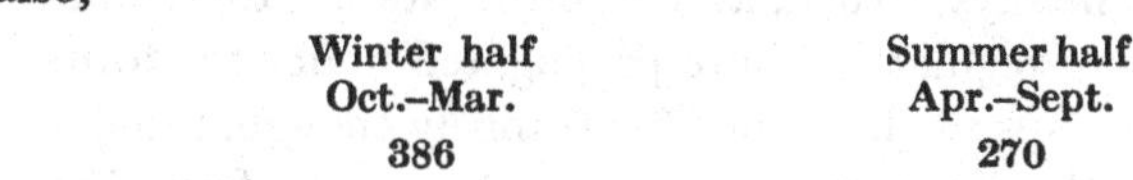

Winter half Oct.–Mar.	Summer half Apr.–Sept.
386	270

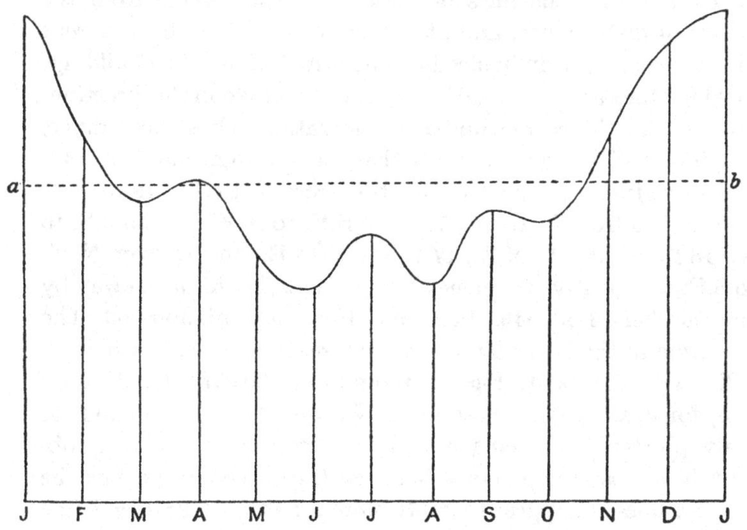

Fig. 2. Seismic curve for France, Belgium and Holland.

the ratio of the latter numbers being as 1 to ·69, or very nearly as 3 to 2; and, lastly,

Winter solst. Dec.–Jan.	Spring equin. Mar.–Apr.	Summer solst. June–July	Autumn equin. Sep.–Oct.
161	83	108	98
1·43	·74	·96	·87

The predominance of earthquakes in the winter and autumn is thus very marked, and it is equally characteristic of nearly every region that Perrey has studied, except the two—Chili and Peru—in the southern hemisphere, in which the inequality disappears or is reversed.

55. In his discussion on the direction of the shock, Perrey was less fortunate. He makes, however, the important observation that, in the epicentral region, there is no prevailing direction to be noticed. "All is irregular, jerking, tumultuous at the centre of disturbance," he says, "then, at a certain distance, the vibrations become more regular, and are propagated under the form of oscillations perhaps isochronous."* Strangely enough, though he realises that the lines of direction must diverge from the centre of disturbance, and at the same time that the apparent directions must be influenced by the orientation of the buildings in which they are perceived†, he groups together all the directions observed in the region under consideration. Thus, for France, Belgium and Holland he finds that, of the total number (149) of observations, 28 give the direction from N. to S., 8 from N.E. to S.W., 35 from E. to W., 11 from S.E. to N.W., 19 from S. to N., 18 from S.W. to N.E., 17 from W. to E., and 13 from N.W. to S.E., the relative frequency being denoted in his usual way by the numbers 1·50, ·43, 1·88, ·59, 1·02, ·96, ·91 and ·69. The variation in the latter figures he represents by a curve which he calls a *seismic rose*. In Fig. 3 is reproduced the curve (continuous line) for France, etc., those for all Europe (broken line) and for Italy (dotted line) being added for comparison. There is obviously a close resemblance between the three curves, but the likeness does not appear to imply more than the natural tendency of observers to refer the direction to one of the four cardinal points‡.

Still less can there be any physical meaning in the resultant direction calculated by Perrey for each of his regions. This is found in the same way as the resultant of forces proportional to the number of observations in each of the eight directions. From

* Lyon, *Soc. Agr. Ann.* vol. 8, 1845, p. 337.

† Brux. *Ac. Bull.* vol. 1, 1857, pp. 113–114.

‡ Taking the eight European regions only, the number of observations in the cardinal directions is 346, and along the bisecting lines 189, or roughly as 2 to 1.

the numbers for France, Belgium and Holland given above, he finds the direction of the resultant to be from N. 71° E. That for the British Isles is from S. 39° W., and that for the whole of Europe from E. 5° S. If we think of the numerous seismic centres scattered over this country, let alone the whole of the continent, it seems clear that the resultant direction is meaningless, even if we were to assume the accuracy of the observations or to recognise, as Perrey does, that the directions are roughly parallel to the axes of neighbouring river-basins or mountain-ranges.

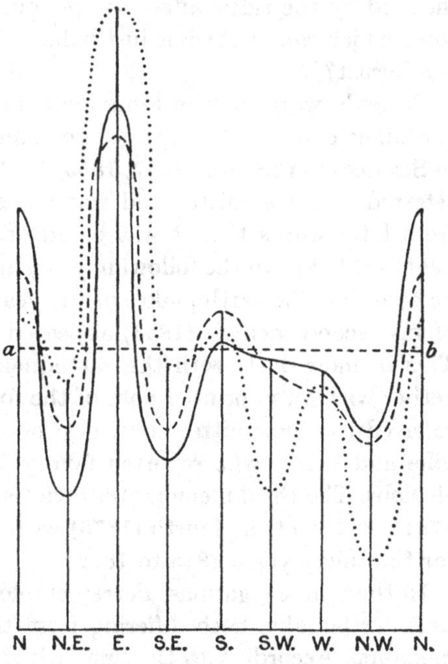

FIG. 3. Seismic roses for France, etc., Europe and Italy.

56. Lunar Periodicity of Earthquakes. Perrey's earliest work on earthquakes, as we have already seen, dealt with their annual variation in frequency. Whatever might be the cause of that variation*, he would naturally inquire if the position of the moon in her orbit could govern a similar variation. That the moon might have some such effect was foreseen as far back as the eighteenth century by G. Baldivi (1669–1707) and G. Toaldo (1719–98)†. Darwin, in 1840, defined the question with his usual insight. "On the hypothesis of the crust of the earth resting on

* He seems to have regarded the variation as due to the earth being in perihelion during the winter months (*Amer. Jl. Sci.* vol. 37, 1864, pp. 1–10).

† G. Baldivi, *Opera Omnia...De Terraemotu Romano, et Urbium adjacentium Anno* 1703, 1737, p. 415; G. Toaldo, *Della vera influenza degli astri*, etc., 3rd ed. 1797, p. 209. Quoted by F. Zantedeschi, Paris, *Ac. Sci. C. R.* vol. 39, 1854, pp. 375–377.

fluid matter, would the influence of the moon," he asks, "as indexed by the tides, affect the periods of the shocks when the force which causes them is just balanced by the resistance of the solid crust?"*

Perrey's work on the lunar periodicity of earthquakes was contained chiefly in four memoirs communicated to the Academy of Sciences in the years 1847, 1853, 1861 and 1875. The first was referred to a committee and was never printed, but we know from later works that it was based on the earthquakes of the years 1801–45. In the following year, in the annual list for 1847, he considers the earthquakes of the years 1844–47. An abstract of his second memoir (1853) appeared in the *Comptes Rendus*. This memoir deals with the earthquakes of 1801–50, and, together with an unprinted note of the following year (1854), was referred to a committee consisting of Élie de Beaumont, Liouville and Lamé, who reported favourably on the author's conclusions. The third memoir (1861) deals with the earthquakes of 1751–1800, and the fourth (1875) with those in the annual lists for the thirty years 1843 to 1872†.

In these investigations, Perrey employs two definitions of the unit earthquake, both differing from that used in the regional memoirs. According to the first, which he soon abandoned, all shocks that occurred on any one day were considered as forming a single earthquake. According to the second, shocks in two, three or four widely separated regions on the same day were counted as two, three or four distinct earthquakes. Thus, Perrey speaks of earthquake-days rather than of earthquakes. He investigates variations in the frequency of earthquakes with regard to (i) the age of the moon, that is, the relative directions of the sun and moon from the earth, (ii) the distance of the moon from the earth, and (iii) the passage of the moon over the meridian. The variations in frequency under the first and second headings are considered in every memoir and in the list for 1847; those under the third heading only in the unprinted note of 1854 and the third memoir (1861).

* *More Letters of Charles Darwin*, vol. 2, 1903, p. 115.

† Dijon, *Ac. Sci. Mém.* 1847–48 (pt. 2), pp. 107–112; Paris, *Ac. Sci. C. R.* vol. 36, 1853, pp. 537–540; vol. 38, 1854, pp. 1038–1046 (report of committee); vol. 52, 1861, pp. 146–151; vol. 81, 1875, pp. 690–692.

57. (i) In his various attempts to discover the relation between earthquake-frequency and the age of the moon, Perrey divides the lunar month of 29·531 days into twelve (and also into eight and six) equal parts, and allots to each part its appropriate number of earthquake-days. For instance, in dealing with the earthquakes of 1844–47, he obtains the following numbers during successive twelfths of a lunation beginning with new moon:

41, 37, 35, 31, 27, 39, 36, 40, 39, 30, 29, 37;

showing an increase in frequency towards the syzygies (new and full moon) and a decrease after the quadratures (first and last quarters). Submitting the same numbers to harmonic analysis, he finds the frequency (number of earthquake-days in one-twelfth of a lunation) represented by the expression

$$35{\cdot}2 - 0{\cdot}7 \sin (t + 4^\circ) + 5{\cdot}1 \sin (2t + 58^\circ),$$

t being an angle measured from the middle of the first twelfth of a lunation*.

The first variable term, which has a period of one lunar month, is thus of little consequence. The second variable term, which has a period of half a lunar month, is more important. Its maxima occur about two days after the syzygies and its minima about two days after the quadratures, the average frequency for one-twelfth of a lunation being about 40 earthquake-days after the syzygies and about 30 earthquake-days after the quadratures.

In his last memoir, Perrey sums up his results in the following table. For this, the lunations are divided into eight equal parts beginning with new moon. The figures in the second column give

	Number of days of earthquakes		
Interval	at syzygies	at quadratures	difference
1751–1800	1904	1754	150
1801–1850	3435	3161	274
1843–1872	8838	8411	427
1843–1847	850	754	96
1848–1852	1054	995	59
1853–1857	1534	1484	50
1858–1862	1603	1537	66
1863–1867	1463	1382	81
1868–1872	2333	2260	73

* Here and elsewhere in this section I have omitted unnecessary decimal places and have given the angles to the nearest degree.

the numbers of earthquake-days in the two-eighths on either side of new and full moon, those in the third column the numbers in the two-eighths on either side of first and last quarters. In the last six lines are given the figures for the six five-yearly periods composing the interval (1843–72) of the annual lists.

Thus, Perrey arrives at his first law that *earthquakes are more frequent at the syzygies than at the quadratures.*

58. (ii) Proceeding in the next place to the influence of the moon alone, Perrey considers the effect of its varying distance from the earth on the frequency of earthquakes. He counts the number of earthquake-days in two intervals of five days each, one consisting of the day of perigee and the two days on either side, the other of the day of apogee and two days on either side. The results are given in the following table:

	Number of days of earthquakes		
Interval	at perigee	at apogee	difference
1751–1800	526	465	61
1801–1850	1223	1113	110
1843–1872	3290	3015	275

Omitting the extreme days of both intervals, the numbers of earthquake-days at perigee and apogee are respectively 314 and 278, 734 and 661, and 1979 and 1839. Perrey's second law is thus that *earthquakes are more frequent at perigee than at apogee.*

59. (iii) In the third line of inquiry, Perrey required long lists of earthquakes recorded at one place of observation, and he counted each shock in the lists as a distinct earthquake. For the first half of the nineteenth century, he made use of de Castelnau's register of 824 shocks at Arequipa from 1810 to 1845. He divided the mean lunar day, of duration 24h. 50m. 30s., into eight equal parts and counted the numbers of shocks in each eighth part. The particular numbers are not given, for the note for 1854 was never printed, but they show that, during the lunar day, there were two epochs of maximum frequency when the moon was close to the upper and lower meridians, and two epochs of minimum frequency when the moon was about 90° from the meridian. For the preceding half-century (1751–1800), he obtained four records of the Calabrian earthquakes of 1783 and following years at Monteleone, Messina, Catanzaro and Scilla, and Arcovito's record at Reggio (Calabria):

Place	Number of shocks when the distance of the moon from the meridian was		
	less than 90°	more than 90°	difference
Monteleone	475	453	22
Messina	84	60	24
Catanzaro	102	81	21
Scilla	140	120	20
Reggio	413	347	66

Thus, Perrey obtains his third law that *earthquakes are more frequent when the moon is near the meridian than when at a distance of 90° from it.*

60. Perrey's confidence in the truth of his three laws was such that they must, he held, be taken into account in any rational theory of earthquakes*. The favourable reception given to them by his contemporaries was not, however, continued by the next generation. An important objection—and one that did not escape the attention of Élie de Beaumont and his colleagues—is the comparatively small number of earthquakes at his disposal, at that time (1854) 6596, though twenty years later it was nearly trebled. A second argument that has been urged, that the variations traced are insignificant, is of little force. Of much more consequence is an objection brought by Montessus (art. 176) against the third law especially, though he holds it to apply also to the others. He divides the lunar day into eight equal parts and finds the ratio of the difference between the greatest and least numbers of earthquakes in each eighth to the total number. If the law were true, this ratio, as the number of earthquakes is increased, should tend to a definite limit; if the law has no foundation, it should tend towards, and ultimately become, zero. The earthquakes studied by Montessus are arranged in seven different classes according to the nature of the record, the numbers of earthquakes in them ranging from 1796 to 15,613, the corresponding values of the ratio being 14·9, 2·9, 2·0, 2·1,

* His own theory is given in a pamphlet of 36 pp. 8vo entitled *Propositions sur les tremblements de terre et les volcans* (Paris, 1863). The part relating to earthquakes is translated and reproduced in *Amer. Jl. Sci.* vol. 37, 1864, pp. 1–10.

1·0, 1·4 and 0·9 per cent. For the total number of earthquakes (44,855) it is 0·66. Thus, Montessus concludes that the frequency of earthquakes has no relation with the culminations of the moon*.

If we apply the same method to Perrey's own tables, the objection loses some of its force. The table given above for the first law does not show this continual decrease, the values of the ratio for totals ranging from 1604 to 17,249 being respectively 6·0, 2·9, 2·8, 1·6, 2·1, 4·1, 1·6, 4·2 and 2·5 per cent. The table given for the second law shows a slight decrease without a tendency to zero, the ratios for 991, 2336 and 6305 earthquakes being respectively 6·2, 4·7 and 4·4 per cent. On the other hand, the table given for the third law shows a continual and marked decrease, the ratio for totals ranging from 144 to 928 earthquakes being respectively 16·7, 11·5, 8·9, 8·7 and 2·4 per cent. Montessus' objection thus applies to the third law, but less strongly, if at all, to the first and second laws.

61. Destructive criticism does not always tend to the advancement of science, but a short paper by A. Schuster (1851–) is a valued exception†. He shows that, if the earthquakes of a given district were to occur entirely at random, harmonic analysis applied to the numbers of earthquakes within given equal intervals may give rise to an apparent seismic period, and that, unless the amplitude of the period considered should exceed a certain expectancy—$\sqrt{(\pi/n)}$, where n is the total number of earthquakes—we have no guarantee as to the reality of the period. In practice, owing to the tendency of earthquakes to occur in groups, the calculated amplitude should amount to two or three times the expectancy. Unfortunately, the tables and curves on which Perrey trusted have never been published, except in the one case of the earthquakes of 1844–47. Representing the mean frequency for this interval (35·2 earthquake-days for a twelfth of a lunation) by unity, the expression of art. 57 becomes

$$1 - \cdot 020 \sin (t + 4^\circ) + \cdot 145 \sin (2t + 58^\circ).$$

* *Arch. Sci. Phys. Nat.* vol. 22, 1889, pp. 422–425. See also the same author's *Tremblements de Terre et Éruptions Volcaniques au Centre-Amérique*, Dijon, 1888, pp. 12–18, and *La Science Séismologique*, 1907, pp. 253–257.

† *Roy. Soc. Proc.* vol. 61, 1897, pp. 455–465.

The number of earthquakes considered being 431, the value of the expectancy is ·085*. As the amplitude of the first variable term is only one-fourth of this, and that of the second nearly double, it follows that the former may be, but is not of necessity accidental, while there is some, though not convincing, reason for believing in the reality of the latter. Thus, to sum up, while Perrey's third law can hardly be maintained, it is possible that his first and second laws may be true, although the foundation that he laid for them is none too secure.

62. Bibliography of Seismology. The earliest attempt to compile a list of memoirs on earthquakes is that of Thomas Young (1773–1829), professor of natural philosophy at the Royal Institution and founder of the undulatory theory of light. The greater part of the second volume of his *Lectures on Natural Philosophy*, etc.,† consists of *A Catalogue of Works relating to Natural Philosophy and the Mechanical Arts*. Under the heading "Meteorology," there is a section on "Earthquakes and Agitations, in order of time" (pp. 490–493), which contains 120 entries, many of them referring to papers in the *Philosophical Transactions*. The chief interest of Young's list at the present time lies in the assignment of certain anonymously written works to definite authors, as, for example, *The History and Philosophy of Earthquakes* to John Bevis (art. 3).

When his second report on earthquakes was published, that is, in 1851, Mallet intended, as he remarks‡, to prepare a complete bibliography of earthquakes. He desisted, however, on learning that Perrey had had such a work in progress for some years. At the time of writing the fourth report (1858), the first part of Perrey's *Bibliographie Séismique* (containing 1836 titles) had appeared. Mallet therefore contented himself by inserting lists of the works on earthquakes to be found in several European libraries, such as the British Museum, the Royal Library at Berlin, the Göttingen university library, etc. The total number of titles in Mallet's list is 435 (four of which are

* It should be remembered that Perrey's definition of the unit-earthquake tends to lessen the effect of clustering.

† 2 vols. London, 1807.

‡ *Brit. Ass. Rep.* 1851, p. 106.

repeated), 279 of them dealing with earthquakes. Many of the memoirs in these lists are little known and difficult of access.

Perrey's *Bibliographie Séismique** is thus the first serious attempt to catalogue the literature of earthquakes. The title, however, is not quite accurate, for the work is, or seems to be, rather a catalogue of the memoirs then in Perrey's library. Only about one-third of the entries actually relate to earthquakes. The titles as a rule are numbered—from 1 to 4015—but, after the numbers were allotted, additions were evidently made to his library, and, apparently to save the labour of re-numbering, these were inserted in their proper places, an asterisk being prefixed in order to show that they are distinct entries. The total number of memoirs catalogued is thus raised to 4098. Of these, 1344 deal with earthquakes, 984 with volcanoes and volcanic eruptions, and 1770 with geology, travels, the heat of the earth's interior, and, in a few cases, with astronomy and meteorology. Of the seismological papers, 535 are written in French, 284 in English, 225 in Italian, 138 in German, and 162 in other languages. Figures such as these, and especially the preponderance of French memoirs, point to the incompleteness of the bibliography, which, nevertheless, contains the titles of many memoirs that are unknown, or but little known, to seismologists of the present day.

63. Conclusion. To sum up, though Perrey's was a well-known name in the third quarter of the last century, he has left no great mark, such as Mallet and Milne have left, on the progress of seismology. The new terms that he invented failed to appeal to his fellow-workers and are now forgotten. His one large view—that of the lunar periodicities of earthquakes—has been a subject of controversy for eighty years, and has not even yet emerged as an established fact. There remain his great catalogues—the annual lists and the regional memoirs—and these surely form an enduring monument of what may be accomplished by one who "seeks a little thing to do, sees it and does it."

* Dijon, *Ac. Sci. Mém.* vol. 4, 1856, pp. 1–112; vol. 5, 1857, pp. 183–253; vol. 9, 1861, pp. 87–192; vol. 10, 1862, pp. 1–53; vol. 13, 1865, pp. 33–102.

CHAPTER V

ROBERT MALLET AND HIS SUCCESSORS

64. To a great extent, the work of Robert Mallet (1810–81) in England was concurrent with that of Alexis Perrey (1807–82) in France, Perrey's lasting from 1841 to 1875 and Mallet's from 1845 to 1878. Perrey's main contributions, as we have seen, were his annual and regional catalogues of earthquakes. In other respects, the ground was unoccupied. Our knowledge of individual earthquakes was gradually increasing, the facts of earthquake-phenomena were becoming known, but, as yet, there was no science of seismology. The next step was made, not by a trained student, but by a busy practical engineer; and, when we think of the work that Mallet accomplished for seismology, we should also remember that it was done, for the most part alone and unaided, in the midst of great engineering enterprises.

Robert Mallet was born in Dublin on 3 June 1810, his father, John Mallet, being the proprietor of a factory in which sanitary fittings and small fire-engines were made. At the age of 16, Mallet entered Trinity College, Dublin, taking the degree of B.A. after a four years' course. In the following year (1831) he was taken into partnership in his father's factory. This he soon converted into engineering works with a considerable foundry, and, before many years had passed, all the engineering work of any consequence in Ireland was carried out by the firm. Mallet's first great feat was to raise the roof of St George's Church, Dublin, a mass 133 tons in weight and covering a large area. This was followed by the building of several bridges over the Shannon, and the invention and construction of hydraulic rams and of ventilating and heating apparatus for public buildings. His buckled plates, which he began to make in 1840, form one of the best floors ever designed, combining the maximum of strength with the minimum of depth and weight. They are to be found on the Westminster and other London bridges. In 1845–46, he was erecting a railway station while he was working at his first important paper on the dynamics of earthquakes. In 1849–55, he built various station roofs and the Fastnet Rock lighthouse, and was designing heavy guns and large mortars, the latter capable of throwing 36-inch shells. At the same time, he was writing his British Association reports on the facts of earthquake-phenomena, determining the velocity of earth-waves in sand and solid rock, and compiling his catalogue of recorded earthquakes. In 1859, he wrote a valuable paper on the coefficients of elasticity and rupture in wrought iron, and worked out the results of his investigation of the Neapolitan earthquake of 1857, and made further experiments on the velocity of earth-waves at Holyhead.

In 1861, engineering work becoming scarce in Ireland, Mallet moved to London, where he set up as a consulting engineer and edited the *Practical Mechanic's Journal*. After an active and healthy life, his eyes began to fail during the winter of 1871–72, and he was practically, though not quite, blind during the last seven years of his life. He died on 5 Nov. 1881.

In scientific thought, Robert Mallet was remarkable for the originality of his ideas, and for the broad grasp he took of every subject that engaged his attention; in private and social life he was beloved for the kindness, geniality and humour of his disposition, for his readiness in conversation and uniform good temper*.

Testimony such as this more than balances the evidence of certain defects that one could have wished to be absent from his writings—his insistent claims to priority and his inaccuracy in referring to the work of other men.

Throughout his life, Mallet's wide interests were retained. Thus, of his papers, which fall into five groups, those on the application of physics to engineering problems date from 1835 to 1875, those on miscellaneous subjects from 1835 to 1874, those on physical geology from 1836 to 1879, those on earthquakes, by no means the most numerous, from 1845 to 1877, while those on volcanic energy and volcanic phenomena came later in his life, from 1862 to 1877. Looking at the list of his papers in chronological order, there is to be seen none of that early groping, as it were, for the main pursuit of his life, with steady concentration on it to the end, though he was occupied chiefly with earthquakes

* *Roy. Soc. Proc.* vol. 33, 1882, p. xx; *Engineering*, vol. 52, 1881, pp. 352–353, 371–372, 389–390; *Inst. Civ. Eng. Proc.* vol. 68, 1882, pp. 297–304.

from 1845 to 1862, and with volcanic energy afterwards. The memoirs and reports on earthquakes, with which we are here concerned, may be divided into six groups dealing with: (i) the dynamics of earthquakes, (ii) the experiments on the velocity of earth-waves, (iii) the catalogue of recorded earthquakes, (iv) the seismic map of the world, (v) the development of the methods of investigating earthquakes, and (vi) their application to the study of the Neapolitan earthquake of 1857.

65. The Dynamics of Earthquakes. Until he was about 35 years old, Mallet, like Michell, had apparently given no thought to earthquakes. His attention was drawn to them, not by a great disaster, but almost by accident. He had noticed in Lyell's *Principles* the well-known diagram (1st ed. 1830, Fig. 21) of a pair of pillars, the upper parts of which had been twisted, without being overthrown, by one of the Calabrian earthquakes. He saw at once the flaw in the then usual statement that, under each pillar, there must have been an independent vorticose movement. His practical training suggested a simple mechanical explanation. If the centre of adherence of the base of the twisted portion were to lie outside the vertical plane containing its centre of gravity and the direction of the motion, the result might be the observed rotation*. The problem is one of minor importance. Its interest lies chiefly in the fact that its solution led Mallet to the views on the nature of earthquake-motion which he described in his memoir on the dynamics of earthquakes†.

In this memoir, in addition to the discussion of the problem referred to, Mallet considers the nature of the earthquake-motion in the earth's crust, the nature and origin of seismic sea-waves and of earthquake-sounds, the need and use of observations on the velocity of earth-waves, and the advantage of founding earthquake-observatories in various parts of the world. There is perhaps little that is really novel in the whole memoir, little, if anything, that the present-day student of seismology would

* Other, and perhaps more probable, explanations of the movement are given in *Japan Seis. Soc. Trans.* vol. 1, pt. 2, 1880, pp. 31–35, and *Geol. Mag.* vol. 9, 1882, p. 264.

† *Irish Ac. Trans.* vol. 21, 1848, pp. 51–105.

have occasion to consult, and his account of Michell's work is strangely inaccurate and incomplete. Its chief merit is that it does form an attempt to explain the more important phenomena of earthquakes by the light of one guiding principle, and thus, as he says, to bring them within the range of exact science. As W. Hopkins (1793–1866) wrote in the following year, he treated the subject "in a more determinate manner and in more detail than any preceding writer."

Others before Mallet had naturally attributed earthquakes to the passage of waves of vibration: for example, J. Michell in 1760 (arts. 16, 18), Thomas Young (1773–1829) in 1807, Gay-Lussac (1778–1850) in 1823, D. Milne (1805–90; art. 44) in 1841, and A. von Humboldt (1769–1859; art. 46) in 1845*. Mallet, however, added precision to previous statements in defining an earthquake as "the transit of a wave of elastic compression in any direction, from vertically upwards to horizontally, in any azimuth, through the surface and crust of the earth, from any centre of impulse, or from more than one, and which may be attended with tidal and sound-waves dependent upon the former, and upon circumstances of position as to sea and land." At the same time, he realised that the waves of vibration may "become complicated by movements of permanent elevation or depression in the land. . . the effects of which it may often be difficult or impossible subsequently to separate."

Mallet gives a very full description of the sea-waves which accompany earthquakes. Although he does not follow Michell in separating the visible waves from the waves of vibration on land, he distinguishes two kinds of sea-waves—the forced sea-wave, as he calls it, and the great sea-wave. His explanations of the origin of both are partially incorrect. The forced sea-wave, which accompanies the earth-wave and is "carried upon its back, as it were," is merely the wave of condensational vibrations in water. If the great sea-wave, which sweeps in some time later, were initially raised by the earth-wave, there would be a great

* *Phil. Trans.* vol. 51, 1761, pp. 571–573; T. Young, *Lect. on Nat. Phil.* vol. 1, 1807, p. 717; *Ann. Chem.* vol. 22, 1823, pp. 428–429; *Edin. New. Phil. Jl.* vol. 31, 1841, pp. 262, 275–277; A. von Humboldt, *Cosmos* (trans.), vol. 1, 1849, pp. 199–200.

sea-wave with every strong submarine earthquake. He does, however, notice that the sea-wave may become "complicated by movements of permanent elevation or depression in the land," and to this extent his explanation is no doubt correct.

Great stress is laid on the importance of determining the velocity of the earth-wave, not so much for the interest of the question itself, as on account of its geological applications. Mallet was aware that the velocity in any rock must depend on the elastic constants of the rock. If we were to measure these constants for all the different rocks, the knowledge of the velocity in a given submarine earthquake, he suggested, would enable us to predict the nature of the underlying rock-formations —an anticipation which has not, I am afraid, been fulfilled.

In his early views on the origin of earthquakes, Mallet was in advance of his time. He connects earthquakes with "local elevations of portions of the earth's crust, often attended with dislocation and fracture of the crust, and sometimes attended with the actual outpouring of liquid matter from beneath." He imagines an earthquake to be produced "either by the sudden flexure and constraint of the elastic materials forming a portion of the earth's crust, or by the sudden relief of this constraint by withdrawal of the force, or by their giving way, and becoming fractured." Where should we look for such sudden changes? Not in volcanic countries, he thinks, but rather in that broad suboceanic belt within which the deposits of the land are accumulated and from which they may afterwards be swept away by tidal currents. "Such a condition of the sea-bottom would seem to be the most likely state of things to give rise to frequent and sudden local elevations or even submarine eruptions of molten matter." And he notices that, while earthquakes are frequent in volcanic countries, they are never of the greatest violence. On the other hand, "the centre of disturbance of almost all the greater earthquakes appears to be beneath the sea, and at considerable distances from active volcanoes," while "the circumstances of the great sea-wave seem to indicate that the centre of disturbance is seldom, if ever, *very* distant from the land."*

* In a later paper (*Geol. Soc. Quart. Jl.* vol. 28, 1872, p. 270), Mallet remarks that "the blow or impulse originating earthquakes could not be

I have given a somewhat full account of this important memoir, for, whatever its imperfections may be, it must, I think, be regarded as one of the chief foundation-stones of seismology as a science. Fortunately, Mallet did not rest content with this, practically his first, contribution. He became afterwards widely known to the scientific public by his four reports to the British Association on the facts of earthquake-phenomena, their publication covering most of the years from 1850 to 1858*. His crowning work on the Neapolitan earthquake of 1857 was published in 1862. With this report and with his memoir on the velocity of earth-waves at Holyhead, his career as a seismologist practically ended, for his study on volcanic energy lies outside our range, and later work, owing to the condition of his eyes, became impossible.

66. Velocity of Earth-Waves. As already remarked, Mallet attached great importance to accurate determinations of the velocity of earth-waves. His first experiments were made in two widely differing materials—the wet sand of Killiney Bay, on the coast of Co. Dublin, and the granite of the neighbouring Dalkey Island. In Killiney Bay, charges of 25 lb. of gunpowder were exploded in the sand, and the passage of the waves at a distance of half a mile was detected by the tremors on the surface of a mercury bath. In Dalkey Island, the ranges were of less than half this length (1021–1155 feet), the granite in one set being of a more fissured or shattered character than that in the other. The mean of eight measurements in sand, six in the fissured

attributed solely to one cause. It arose often from deep subterranean volcanic action; but it also—especially in the case of long-continued tremors, like those of Comrie or Pignerol—arose from the breaking up or the grinding over each other of rocky beds at a great depth, through the tangential pressures produced in the earth's crust by secular cooling."

* Of the four reports, the first (1850) deals with the facts of earthquake-phenomena and ancient and modern theories of their origin, and is remarkable for its evidence of wide and varied reading; the second (1851) describes his experiments on the velocity of earth-waves and the construction of his great catalogue; the third (1852–54) consists of the great catalogue; and the fourth (1858) contains his discussion of that catalogue, a summary of the work of Alexis Perrey, the lists of works on earthquakes that he had found in several libraries, etc.

granite and three in the other, gave the following velocities after all corrections were applied:

Sand	825	feet per second
Discontinuous granite	1206	,,
More solid granite	1665	,,

The second series of measurements were made at Holyhead from 1856 to 1861, large masses of rock being dislodged from the Government quarries by the explosion of from one to five or more tons of gunpowder in mines. The ranges (six in number) varied from slightly less than a mile to about a mile and a quarter, crossing different lengths of quartz rock, slates and schists. The velocities varied from 954 to 1289 feet per second, the average of all six measurements being 1089 feet per second*.

67. Catalogue of Recorded Earthquakes. In the compilation of his earthquake-catalogue, as well as in the experiments described above, Mallet was assisted by his eldest son, John William Mallet (1832–1912), afterwards professor of chemistry in the university of Virginia. The catalogue forms his third report on earthquake-phenomena and occupies nearly 600 pages in the reports of the British Association (1852–54). The discussion of the catalogue and the description of the seismic map of the world are contained in the fourth report (1858)†.

Though based on several earlier catalogues, and especially on those of Hoff and Perrey, Mallet regarded his catalogue as, in his belief, the "first attempt to complete a catalogue that shall embrace all recorded earthquakes." This statement, however, is hardly just to Hoff, whose *Chronik der Erdbeben und Vulkan-Ausbrüche*, if less full than Mallet's catalogue, is a very valuable work (art. 41).

* *Brit. Ass. Rep.* 1851, pp. 272–317; *Phil. Trans.* 1861, pp. 655–679; 1862, pp. 663–676. The above velocities are given to the nearest foot per second. Mallet expresses them to three places of decimals. Such detail, however, is meaningless, for an error of one-hundredth of a second in the time of transit at Killiney Bay involves an error of more than two feet per second in the velocity. In two of the Holyhead experiments, the charges fired were not very different (2100 and 2600 lb.). As the distances traversed in quartz rock and schist rock are known for both ranges, we find the velocity in quartz rock to be 1162 feet per second and in schist rock 1075 feet per second.

† *Brit. Ass. Rep.* 1852, pp. 1–176; 1853, pp. 118–212; 1854, pp. 1–326; 1858, pp. 1–136.

The labour involved in the preparation of Mallet's catalogue must have been immense. For about one-half the total number of entries, he was indebted to the catalogues of other writers, such as É. Bertrand, L. Cotte, P. Merian, etc.* The other half were obtained from scattered sources—books of travel, and British, French and German newspapers and scientific journals. Mallet was not content with copying from the earlier catalogues, but, in every possible case, consulted the original author or authors referred to by his predecessors. The details relating to each earthquake are given in tabular form under the headings (i) time of occurrence, (ii) area chiefly affected, (iii) the direction, duration and number of the shocks, (iv) the accompanying or following sea-waves, (v) meteorological and secondary phenomena, and (vi) the authorities for the records. The first entry is under the date 1606 B.C., and Mallet proposed that the catalogue should end with the year 1850. It was found, however, that the completeness of Perrey's annual lists rendered its extension beyond the year 1842 unnecessary, but the earthquakes of the eight omitted years are included in the discussion in the fourth report. The total number of earthquakes recorded is stated by Mallet to be 6831, and, taking into account this great number and also the scantiness of the materials available, he felt himself justified in stating his conviction "that nearly all that can be drawn from the collection and discussion of such records has now been done." Mallet may have erred in regarding his catalogue as having no forerunners. There can be little doubt, however, that it is the last that will ever be published on so extensive a scale.

Of the earthquakes included in this valuable list, 216 were "great," or strong enough to reduce whole towns to ruins. Judging from the last 150 years, during which the record of such disasters may be supposed complete, Mallet estimates that a great earthquake occurs on an average once every eight months. The rapid expansion in the number of entries is no proof, in Mallet's opinion, of any actual increase in earthquake-frequency.

* É. Bertrand, *Mémoires Historiques et Physiques sur les Tremblemens de Terre*, La Haye, 1757, 328 pp.; *Jl. Phys.* vol. 65, 1805, pp. 161–168, 250–264, 329–364; P. Merian, *Ueber die in Basel wahrgenommenen Erdbeben*, Basel, 1834, 20 pp.

He regards it as, in fact, "a record of the advance of human enterprise, travel, and observation," the evidence tending rather to the conclusion that "during all historic time the amounts of seismic energy over the observed portions of our globe must have been nearly constant." At the same time, there is clear evidence of irregular and paroxysmal outbursts of energy in reference to shorter periods. The frequency curves for the last three centuries and a half (1501–1850) show, indeed, that, while the least interval of repose may be a year or two, the average interval is from five to ten years; that the shorter intervals are usually connected with periods of diminished earthquake frequency; that the alternations of paroxysm and repose appear to follow no discernible law, except that two marked periods of extreme paroxysm occur in each century, the greater about the middle and the less not far from the end.

68. Seismic Map of the World. In his seismic map of the world, Mallet was breaking almost new ground, though two similar attempts had been made during the preceding fifteen years. The *Physikalischer Atlas* of Heinrich Berghaus (1797–1884), published in 1845, contains a chart of the volcanic phenomena of the Old World* in which the regions disturbed by earthquakes are indicated by shading, deeper in proportion to the intensity and frequency of the shocks. Curves are also drawn bounding the areas affected by three remarkable earthquakes—the Lisbon earthquake of 1755, the Caraccas earthquake of 1812, and the South-east Europe earthquake of 1838. The map is based on Hoff's complete work, of which the *Chronik der Erdbeben* forms the fourth and fifth volumes. Though it is said to represent the Old World, it includes the whole of South America, most of Central America and the eastern half of North America, and on the other hand omits the east of Asia and its important island districts. The *Physical Atlas* of Alexander Keith Johnston (1804–71), published in 1848, contains a similar map of the whole world†, based on that of Berghaus and on the work of Hoff. In this, the seismic districts are tinted in the same

* Gotha, 1845, Sect. 3, *Geologie*, no. 7.
† Edinburgh, 1848, *Geological Division*, no. 7.

way as in Berghaus' map, and curves bound the disturbed areas of six important earthquakes. The great seismic zone running from the Aleutian and Kurile Islands through Japan and the Philippines to the East Indies is clearly marked, but, as in Berghaus' map, there are many regions that are unavoidably left blank*.

69. Mallet's materials, though far from complete, were much fuller than those at the disposal of Berghaus and Johnston. His original map, on Mercator's projection, is 75 inches long and 48 inches wide and is now in the possession of the Royal Society. In his fourth report, this map is reduced to about one-third linear dimensions†. He divided earthquakes into three classes—great, mean and minor. If their disturbed areas were unknown, he assumed that their radii were 540, 180 and 60 geographical miles, and their areas were coloured by three tints, the intensities of which were as the numbers 9, 3 and 1. Thus, the most deeply shaded areas on the map represent on a definite scale those in which earthquakes are most frequent and violent.

While its defects are obvious, being those of all maps in which disturbed areas are depicted, Mallet's map remained for nearly half a century our best representation of the distribution of earthquakes over the globe. The more important results which he deduced from it are the following: as the distribution of earthquakes is paroxysmal in time, so also it is local in space; the normal type of distribution is in bands of variable and great width (from five to fifteen degrees); these bands generally follow the lines of elevation which divide the great oceanic or terr-oceanic basins of the earth's surface; in so far as these are frequently the lines of mountain-chains, and these latter of volcanic vents, so the seismic bands are found to follow them likewise; the regions of least or no disturbance are the central areas of great oceanic or terr-oceanic basins and the greater islands existing in shallow seas. The modern and more accurate method of mapping by epicentres, rather than by disturbed areas, has led to greater detail in our knowledge, but the main

* In the second edition of 1856 (vol. 1, plate 10), most of the blank regions remain, though seven new curves are given bounding the disturbed areas of earthquakes in northern India from 1819 to 1852.

† *Brit. Ass. Rep.* 1858, plate 12.

laws of seismic distribution are those which Mallet has so clearly established.

70. Methods of Investigation. Though they are not explicitly stated, there can be little doubt, I think, that the methods of investigation that Mallet afterwards developed were partly present in his mind when he wrote his early papers. Even in the first edition (1849) of the *Admiralty Manual of Scientific Enquiry*, to which, on Darwin's recommendation, he was invited to contribute, there is no direct account of them in the section on the observation of earthquake-phenomena. They are described for the first time, and very fully described, in his report on the Neapolitan earthquake, and it is some indication of the importance that he attached to them that the main title given to these two large volumes is *The First Principles of Observational Seismology**.

The fundamental object of Mallet's inquiry was to ascertain the surface-position and depth of the seismic focus. The methods that he proposed for this purpose are well known. By observations on the direction of fissures in large and uniform buildings, of the fall of columns, and of the projection of detached masses of masonry, he sought to determine the horizontal direction of the shock, and such directions at two or more places would suffice to find the position of the point vertically above the seismic focus, now known as the epicentre. Once this point is determined, a single observation of the angle of emergence, as given by the inclination of fractures in walls, would lead to the depth of, at any rate, one point within the focus.

Various writers before Mallet had suggested the use of observations on the direction of the shock. The earliest case is probably that of J. Michell in 1760 (art. 22). D. Milne used the method in 1841 to determine the epicentre of the Comrie earthquakes so frequent at that time, the directions at different places being obtained chiefly from the records of inverted pendulums (art. 44). Lastly, W. Hopkins (1847) gave the same method for locating the epicentre (art. 80)†. It is one thing, however, to suggest a method, and quite another to apply it; and few, I think, will be

* Vol. 1, 1862, pp. 5–160.

† *Phil. Trans.* vol. 51, 1761, pp. 625–626; *Brit. Ass. Rep.* 1841, p. 48; 1842, pp. 96–97; 1843, p. 121; 1847, pp. 82–83.

found to deny that, if any name is to be associated with these methods, it should be that of Robert Mallet.

How difficult the application of the method would prove to be, Mallet did not realise until he was immersed in the investigation of the Neapolitan earthquake.

> "When the observer," he says, "first enters upon one of those earthquake-shaken towns, he finds himself in the midst of utter confusion. The eye is bewildered by 'a city become an heap.' He wanders over masses of dislocated stone and mortar....Houses seem to have been precipitated to the ground in every direction of azimuth. There seems no governing law, nor any indication of a prevailing direction of overturning force. It is only by first gaining some commanding point, whence a general view over the whole field of ruin can be had, and observing its places of greatest and least destruction, and then by patient examination, compass in hand, of many details of overthrow, house by house and street by street, analysing each detail and comparing the results, as to the direction of force, that must have produced each particular fall, with those previously observed and compared, that we at length perceive, once for all, that this apparent confusion is but superficial."

It would seem from this that Mallet, in part perhaps unconsciously, selected the directions which appeared from such a survey to predominate. In other words, he obtained by inspection a rough estimate of the average direction of fall, and, as Omori showed in 1894, the average of a large number of observed directions, even very various directions, may give a very close approximation to the true direction of motion*.

71. From 1849 to 1871, Mallet contributed the article on the observation of earthquake-phenomena to four editions of the *Admiralty Manual of Scientific Enquiry*†. The article in the fourth edition, almost his last writing on earthquakes, contains a summary of his previous work—the phenomena as described in his British Association reports and the methods of investigation that he had tested in the kingdom of Naples. He also suggests the use of simple apparatus, of an instrument to give the time of an earthquake by catching the pendulum of a clock,

* *Neapolitan Earthquake of* 1857, vol. 1, pp. 35–36. During the Tokyo earthquake of 20 June 1894, Omori found the mean direction of fall of 152 stone lanterns in Tokyo to be S. 71° W., almost exactly the same as the direction of the principal vibration of that earthquake, namely, S. 70° W. (arts. 220, 230).

† 1849, pp. 196–223; 1851, pp. 205–236; 1859, pp. 325–363; 1871, pp. 299–333.

and of two series of six cylinders each erected along perpendicular lines, the diameters of the cylinders ranging from one-ninth to one-third their height. Cylinders of this type were erected in 1874 in the "earthquake-house" at Comrie and have also been tested in Japan, in neither case with satisfactory results*.

72. Study of the Neapolitan Earthquake of 1857. Owing to his long residence in Ireland—a typical aseismic country—Mallet had no personal acquaintance with earthquakes until 1852. On 9 Nov. of that year he was awakened by a strong shock, one that is perhaps unique among British earthquakes in having affected all four portions of the United Kingdom. It failed, however, to give him the materials that he required†, and five years more had to pass before they were provided by the destructive Neapolitan earthquake of 16 Dec. 1857. Aided by a grant of £200 from the Royal Society he proceeded to the stricken area. Leaving Naples on the following 10 Feb., he spent about two months visiting the ruined towns and villages of the meizoseismal area, heedless of the many discomforts of a camp life during the wet and cold of winter, convinced that in fractured walls and overthrown pillars he had "the most precious data for determining the velocities and directions of the shocks that produced them," inspired by the thought that for the first time the depth of the seismic focus was being "measured in miles and yards with the certainty that belongs to an ordinary geodetic operation." Sixty years and more have passed since then; the certainty of 1858 may have given way to doubt; yet, as one reads the account of his work at Polla and Vietri di Potenza, when he realised that at last he had found the point vertically above the origin of the shock and could state, as a first approximation, that the depth of the focus was 5·64 geographical miles, one cannot but sympathise with his confidence that he was showing the way to a true intelligence of "the viewless and unmeasured miles of matter beneath our feet."

73. Mallet's report on the Neapolitan earthquake is one of the earliest of a long series of monographs on special earthquakes,

* *Brit. Ass. Rep.* 1874, p. 241; 1875, pp. 64–65; *Japan Seis. Soc. Trans.* vol. 3, 1881, pp. 46–52.

† *Irish Ac. Trans.* vol. 22, 1855, pp. 397–410.

and, for forty years, until the Assam earthquake of 1897, it remained without a rival. It was published in two large volumes, and with a wealth of maps and plates that has only been exceeded in the great reports on the Californian earthquake of 1906 and the Japanese earthquake of 1923*. His maps of the earthquake mark an epoch in seismology. Although not for the first time, isoseismal lines are clearly depicted (Fig. 4)†. There are

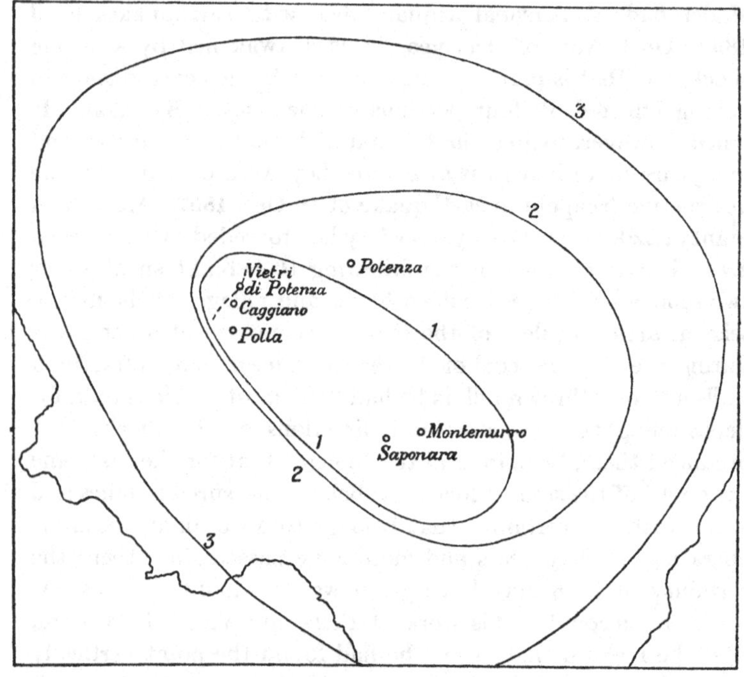

Fig. 4. Neapolitan earthquake of 16 Dec. 1857; map of isoseismal lines.

four of these. The first surrounds the meizoseismal area, within which the towns were for the most part prostrated; the second

* *Great Neapolitan Earthquake of 1857: the First Principles of Observational Seismology*, 1862. The two volumes contain altogether 854 pages, 4 large folding maps, 57 plates and 158 diagrams in the text. The Royal Society contributed £300 towards the cost of publication. For earlier reports on great earthquakes, see arts. 26–32.

† For earlier instances of the construction of isoseismal lines, see arts. 42, 120, 121.

includes places in which large parts were thrown down and persons were killed; the third those in which slight damage to buildings occurred without any loss of life; while the fourth bounds the area within which the shock was perceived without instrumental aid.

74. On this map, Mallet also entered, and for the first time*, the lines of wave-path at different places, taking the most probable direction measured or the mean of the extremes when all observations seemed equally correct. Altogether he made 177 observations on the direction at 78 places. Of these, 16 lines pass through a point close to the village of Caggiano or within a distance of 500 yards from it; 32 pass within a distance of 1·2 miles† from the same point, 16 others within a distance of 2·9 miles, and 12 within a distance of 5·8 miles. The evidence from the first group alone "is irresistible," says Mallet, "that we have obtained the real position of the place upon the earth's surface vertically above that one beneath, whence the shock emanated." He attributes the slight divergence of the remaining observations chiefly to the dimensions of the focus, partly also to errors of observation and drawing, and to the reflection and refraction of the earth-waves. In fact, he uses the observed divergences as evidences of the horizontal length of the focus, which he estimates at about 10½ miles (vol. 2, pp. 235–247).

75. Having determined the position of what we now call the epicentre, he proceeded to find the depth of the focus. For this purpose, he relied on 26 estimates of the angle of emergence made at places within 40 miles from Caggiano. The greatest depth indicated by these angles of emergence is 9·3 miles, and the least 3·2 miles. Of the wave-paths, 18 start from the seismic vertical (or vertical line through the focus) within a range of 2·3 miles, and the mean focal depth, as determined from them, is 6·6 miles. Looking at the points on the seismic vertical (Fig. 5) from which the wave-paths chiefly diverge, Mallet considers that the probable vertical dimension of the focus does not exceed 3·4 miles (vol. 2, pp. 248–251).

* Omitting the attempt on the British earthquake of 1852.

† Mallet in this and other cases used the geographical mile (1·153 statute miles) as his unit of length.

76. Mallet's remaining conclusions may be considered at less length. He was obviously puzzled by the position of the epicentre near one end of the innermost isoseismal, and tried to account for it by the physical structure of the country and the direction of the focus (vol. 2, pp. 274–275)*. The mean surface-velocity of the earth-waves he estimated at the rather low value of 788 feet per second (vol. 2, pp. 322–334). The amplitude of the wave-motion he found to range from 2·5 to 4·75 inches at distances between 4·0 and 30·1 miles from Caggiano and to increase with the distance from that village (vol. 2, pp. 347–350). The maximum velocity of a wave-particle was determined from the overthrow or projection of bodies at twelve places between the same distances from Caggiano, and was found to lie between 9·8 and 21·2 feet per second and to diminish as the distance increased. The mean of eleven good observations was 12·4 feet per second (vol. 2, pp. 335–341).

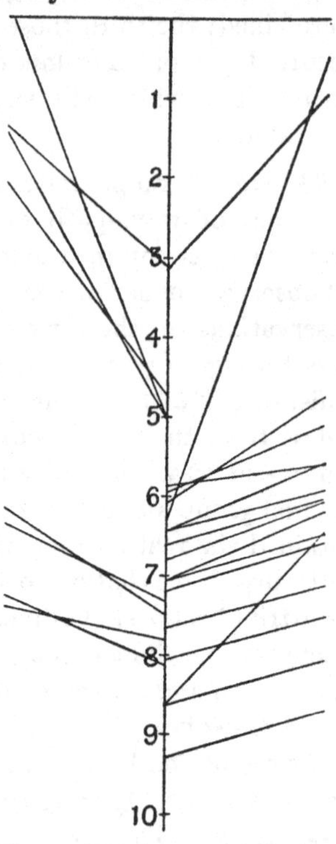

Fig. 5. Neapolitan earthquake of 1857; wave-paths in the neighbourhood of the seismic vertical.

77. Conclusion. To sum up Mallet's contributions to seismology is not an easy task. One of the most important is also the most intangible—his influence on the different points of view from which earthquakes were regarded, say in 1845 and after the lapse of twenty years. Fifteen years after his death and nearly

* A more probable explanation of the excentric position of the epicentre is that there were two foci, one near Caggiano, and the other about 24 miles to the south-east near Montemurro. This double focus would account for the double shock so frequently observed and for the very high death-rates near the assumed south-east epicentre—71 per cent. at Montemurro and 50 per cent. at Saponara.

forty years after the Neapolitan earthquake, I was struck by the fact that, of more than two thousand observers of the Hereford earthquake of 1896, one in every five gave unasked his impression of the direction of the shock. In the large towns, the proportion rose to one in every three.

"It is given to no man," said Mallet, "so to interpret nature that his enunciation of her secrets shall remain for ever unmodified by the labours of his successors." Mallet's work was unfortunately no exception to this rule. Some of it may now be obsolete, but much of it remains. Several of the terms in daily use—seismology, seismic focus, angle of emergence, isoseismal line, and meizoseismal area—are due to him. The position of the epicentre may still be determined by observations on the direction of the shock, especially by the mean of a large number of observations. As regards the depth of the focus, the results derived from such observations are much less certain. All that can be said is that they perhaps indicate the order of magnitude of the true depth. His experiments on the velocity of earth-waves are interesting, though their bearing on the actual problem is somewhat remote. But, by his perception of the nature of earthquake-motion, by the construction of isoseismal lines, by the compilation of his great catalogue, by his statement of some of the laws which govern the distribution of earthquakes in time and space, and, above all, by his investigation—the first really scientific investigation—of a great earthquake, Mallet has placed the science of seismology under a debt, which those who have followed in his steps would be the last to under-estimate.

SUCCESSORS OF ROBERT MALLET

78. Of the three men whose work is referred to in the present section, William Hopkins (1793–1866) was a successor, but in no sense a follower, of Robert Mallet. Thomas Oldham (1816–78) and Henry James Johnston-Lavis (1856–1914) were directly inspired by Mallet's investigation of the Neapolitan earthquake. They followed his methods implicitly, the one in studying the Cachar earthquake of 1869, and the other the Ischian earthquakes of 1881 and 1883, in both cases under difficulties more serious than even he encountered in the kingdom of Naples.

79. William Hopkins (1793–1866) at first followed his father's occupation as a farmer, but, having no taste for the work, he sold his property and entered Peterhouse, Cambridge. After taking his degree (as 7th wrangler) in 1827, he settled in Cambridge as a private tutor, a profession that he followed with such success that he became known as the "senior wrangler maker," Stokes, Kelvin, Tait, Maxwell and Todhunter being among his pupils. Influenced by Sedgwick's enthusiasm, he was led to study geology, and, from 1836 until the last year of his life, he issued a series of memoirs in which he applied mathematical analysis to the solution of geological problems*. He was, for instance, the first to realise the great thickness of the earth's crust, which he thought to be not less than from one-fifth to one-fourth of the radius†.

80. In 1847, Hopkins contributed to the British Association a report on the geological theories of elevation and earthquakes‡. The greater part of it (pp. 40–74) is devoted to a summary of his early researches in physical geology, the second part (pp. 74–92) to earthquakes and earthquake-motion. It should be remembered that Mallet's memoir, to which he makes a cordial reference, was not published in full until the following year. Hopkins, treating the theory as obvious—as it must indeed have seemed to a mathematical physicist at that time—gives an account of wave-motion in liquid and solid bodies that may still be read with advantage.

In applying these principles to earthquakes, Hopkins at first assumes the seismic focus to be of small dimensions, "as would be the case...if the shock were produced by a deep-seated volcanic explosion, the falling-in of the roof of a subterranean

* It is said that he used to complain that he could not get geologists to understand his mathematics nor mathematicians to take an interest in his geology; but his work was nevertheless appreciated by geologists, for he was awarded the Wollaston medal by the Geological Society in 1850 and was president of the Society from 1851 to 1853.

† *Dict. Nat. Biog.* vol. 27, 1891, pp. 339–340; *Geol. Mag.* vol. 3, 1866, p. 576.

‡ *Brit. Ass. Rep.* 1847, pp. 33–92.

cavity, or the sudden rending of the solid rock around it." He then shows how the most important point, the position of the focus, may be determined. That of the epicentre he would find from the intersection of two lines of direction at the surface, but he makes the important remark that the direction observed should be that at the first instant of its motion, for, afterwards, the later vibrations of the condensational waves and the earlier vibrations of the distortional waves may coalesce—a precaution that lies at the root of Galitzin's method. He also points out that the two lines of direction should be nearly at right angles.

Hopkins' methods of determining the depth of the focus are of less practical value. He supposes that the velocities of the condensational and distortional waves (V_1 and V_2) in the solid crust are known, and that the horizontal or surface-velocity (v) at a point distant a from the epicentre can be measured. According to the first method, the depth of the focus would be

$$a\frac{\sqrt{(v^2 - V_1^2)}}{V_1}.$$

The second method requires the measurement at some station of the interval T between the arrivals there of the first condensational and distortional waves, that is, of the duration of the first preliminary tremors. The distance of the focus would then be

$$T\frac{V_1 V_2}{V_1 - V_2}.$$

Knowing the position of the epicentre, the depth of the focus can be calculated.

Hopkins also suggests the use of the last expression for finding the distance of the epicentre of a submarine earthquake from a station on the coast, the interval T to be observed being that between the arrivals of the first vibrations and the sea-waves. This has recently been used by K. Suda to check the position, otherwise determined, of the epicentre of the Japanese earthquake of 1 Sep. 1923*.

These are perhaps the most important points of a memoir so full of interesting remarks that one cannot help regretting that Hopkins' attention should have been diverted from earthquakes

* Kobe (Japan), *Marine Obs. Mem.* vol. 1, 1924, p. 143.

to, say, his lengthy mathematical discussions on the motion of glaciers.

81. One of the earliest of Mallet's successors was Thomas Oldham (1816–78), who was trained in geology under Robert Jameson (1772–1854) at Edinburgh university. In 1845, he succeeded John Phillips (1800–74) as professor of geology in Dublin university, and in the following year was appointed director of the Geological Survey of Ireland. From 1850 to 1876, he was at the head of the Geological Survey of India, and it was towards the close of his service in that country that his interest in seismology was roused by the strong Cachar earthquake of 10 Jan. 1869. With Mallet's Neapolitan report as his guide, he examined as much as was possible in the then disturbed state of the country. On his return to England, he brought with him the materials collected, but the report was still unfinished when he died in 1878*.

82. Oldham's contributions to seismology consist of two memoirs† and neither of them unfortunately did he live to see in print. The account of the Cachar earthquake‡, edited by his son Richard Dixon Oldham (1858–), follows closely on the lines of Mallet's great report, but his materials were much less ample. There were, for instance, only 12 observations of the direction, and 6 of the angle of emergence. The lines of direction intersect within an area 20 to 30 miles long, and 3 or 4 miles wide, on the northern border of the Jaintia Hills (pp. 60–65). The mean depth of the focus was probably about 30 miles, with a possible error of 5 miles on either side (pp. 66–71). The maximum velocity of a wave-particle at the surface he estimated at 20 feet per second (pp. 71–80), and the mean velocity of the earth-wave at 7375 feet per second, a velocity so high that he regards it as improbable (pp. 81–84).

In the following year was published his *Catalogue of Indian*

* *Geol. Soc. Quart. Jl.* vol. 35, 1879, *Proc.* pp. 46–48; *Geol. Mag.* vol. 5, 1878, pp. 382–384.

† Not counting a letter on the secondary effects of the Cachar earthquake which Mallet communicated to the Geological Society (*Geol. Soc. Quart. Jl.* vol. 28, 1872, pp. 257–260).

‡ *India Geol. Surv. Mem.* vol. 19, pt. 1, 1882, pp. 1–98, 15 plates.

earthquakes from the earliest time to the end of A.D. 1869*. The first year mentioned is A.D. 893 or 894, but, up to the end of the sixteenth century, there are only three entries. In the seventeenth century 8 earthquakes are recorded, and in the eighteenth 7, while in the nineteenth century the number mounts up to 334. Two earthquakes are distinguished by great loss of life. In that of 893 or 894, 180,000 persons are reported to have been killed, and 300,000 in that of 1737.

83. Henry James Johnston-Lavis (1856–1914), the student of Vesuvian phenomena and the historian of Ischian earthquakes, received his medical training at University College, London. Here he came under the influence of John Morris (1810–86), professor of geology. From 1879 to 1894, he lived in Naples, where he established a practice among English and American residents, though from 1891 onwards for several years the summer months were spent in Harrogate. The main work of his life was no doubt his survey of Vesuvius made during the years 1880–88, the final map being published in 1891 in six sheets on the scale of 1 : 10,000. As secretary of the British Association committee on the volcanic phenomena of Vesuvius, he wrote the eleven reports presented from 1885 to 1895. He also became professor of vulcanology in the university of Naples in 1892. On leaving that city in 1894, he practised successively at Monte Carlo, Beaulieu (Alpes-Maritimes) and Vittel. Being in Paris shortly after the war broke out, and finding the railways blocked, he started for Vittel by motor-car, from which, owing to an accident, he was thrown and killed on 10 Sep. 1914†.

During the early years of his residence in Naples, on 4 Mar. 1881 and 28 July 1883, two destructive earthquakes of the volcanic type occurred in the neighbouring island of Ischia. On each occasion, he spent about three weeks in the island, measuring the direction and inclination of fissures in buildings. In both cases, the accuracy of the measurements was affected by the irregularity and poor quality of the houses, the existence of

* *India Geol. Surv. Mem.* vol. 19, pt. 3, 1883, pp. 1–53.

† Johnston-Lavis, *Bibliography of the Geology and Eruptive Phenomena of the more important Volcanoes of Southern Italy*, 1918, introd. by B. B. Woodward, pp. vii–xiii; *Geol. Soc. Quart. Jl.* vol. 71, 1915, pp. lviii–lix.

other fissures in them due to previous earthquakes, and the geological structure of the island. The results of Johnston-Lavis' survey were published in several short papers, more fully afterwards in his *Monograph of the Earthquakes of Ischia**. Making use of 55 azimuths and 9 angles of emergence for the earthquake of 1881, and of 69 azimuths and 24 angles of emergence for that of 1883, he found that the earthquakes originated in almost coincident foci, the epicentre being close to Casamenella and the depths of the foci 618 and 628 metres respectively†. The principal outcome of Johnston-Lavis' work is his determination of the small depth of the focus in earthquakes of volcanic origin, a result that was qualitatively evident from the small areas disturbed by earthquakes so destructive at the epicentre.

84. Mallet's method of finding the depth of the focus has been applied by other seismologists, whom one can hardly claim as his followers, with the following (approximate) results: for the Ischian earthquake of 1883, ¾ mile (Mercalli); for the Andalusian earthquake of 1884, 7½ miles (Taramelli and Mercalli); for the Kashmir earthquake of 1885, 7½ miles (Jones); for the Bengal earthquake of 1885, 50 miles (Middlemiss); for the Riviera earthquake of 1887, 10¾ miles (Mercalli); and for the Verny (Turkestan) earthquake of 1887, 6½ miles (Mouchketow)‡.

* London, 1885, 112 pp. 4to, 26 plates.

† A few mathematical calculations were made by Samuel Haughton (1821–97), professor of geology in the university of Dublin, who had provided the corresponding formulae in Mallet's work on the Neapolitan earthquake.

‡ G. Mercalli, *L' isola d' Ischia ed il terremoto del* 28 *luglio* 1883, pp. 133–134; T. Taramelli and G. Mercalli, Roma, *R. Acc. Lincei Mem.* vol. 3, 1886, pp. 69–71; E. J. Jones, *India Geol. Surv. Rec.* vol. 18, 1885, pp. 224–225; C. S. Middlemiss, *ibid.* p. 211; G. Mercalli, Roma, *Uff. Centr. Meteorol. Ann.* vol. 8, 1888, pp. 246–248; J. V. Mouchketow, St Pétersb. *Com. Geol. Mém.* vol. 10, 1890, p. 149.

CHAPTER VI

THE STUDY OF EARTHQUAKES IN ITALY

85. The deep interest roused by the Calabrian earthquakes of 1783 hardly survived the eighteenth century. Partly this may have been due to the fact that successors of equal severity were wanting. During the century that followed them, a violent earthquake occurred every five or six years in Italy, but only three were disastrous—the Molise earthquake of 1805, the Neapolitan earthquake of 1857, and the Ischian earthquake of 1883 (arts. 72–76, 83, 84, 106).

The Tuscan earthquake of 14 August 1846 belonged to a lower rank of intensity, but it was made the subject of a useful report* by Leopoldo Pilla (1805–48), of Pisa, who devoted much of his time to the study of Vesuvius. Pilla's volume is noteworthy on two accounts: (i) the catalogue of Tuscan earthquakes from 991 to 1843, and (ii) the study of the relations between the damage to property and the constitution of the ground. He shows that all the deaths (about 60 in number) occurred in houses built on the recent sub-Apennine formation, that, on Miocene strata, many persons were wounded, while, on the older rocks, not a single building was destroyed.

From the time of Pilla onwards, that is, from the middle of the nineteenth century, the study of Italian earthquakes has grown and never failed—a result for which we are mainly indebted to the work of Luigi Palmieri (1807–96), Timoteo Bertelli (1826–1905), Michele Stefano De Rossi (1834–98), Giuseppe Mercalli (1850–1914), and the band of students who form, or have formed, the Italian Seismological Society.

LUIGI PALMIERI AND THE VESUVIAN OBSERVATORY

86. Luigi Palmieri was born on 21 Oct. 1807 at Faicchio, in the province of Benevento. After studying mathematics in the

* *Istoria del Tremuoto che ha devastato i paesi della costa Toscano il dì 14 Agosto* 1846, Pisa, 1846, 226 pp.

university of Naples, he became a teacher of mathematics and philosophy in schools of that city, and, in 1847, professor of philosophy in the university. In 1860, he was elected to a chair specially created for him, that of terrestrial physics. He died at Naples on 9 Sep. 1896.

Palmieri's name will always be associated with the observatory on the south side of Vesuvius, the building of which was begun in 1841, though not finished until 1847. The first director, Macedonio Melloni (1801–54), celebrated for his researches on radiant heat, was involved in the political disturbances of 1848, and was forced to retire without ever entering into office. Six years later, he died of cholera near Naples. In 1852, Palmieri was allowed to make investigations in the observatory, and, in the first of many papers on Vesuvius, he described the eruption of 1 May 1855, as witnessed from it. In 1856, he was formally appointed director.

Palmieri's chief contribution to seismology was the invention of his electro-magnetic seismograph in 1855. In the following year, one constructed from his designs was placed in the Vesuvian observatory. A second was erected in a building belonging to the university of Naples. Though not the earliest seismograph in action, it was probably the first that recorded tremors imperceptible to human beings, for the Forbes seismometer and other pendulums erected at Comrie in 1840 failed to register many earthquakes that were readily felt in the village*. This, no doubt, was one of the reasons why Palmieri was so highly esteemed by his compatriots. Even at the present day, when seismographs are so common, it is the registration of a distant earthquake, not the havoc wrought by it, that appeals to the public with an unfailing interest.

But the chief reason for his popularity lay in his sterling character, his attractive and graceful mode of speech, his simple and courteous manner, and his utter freedom from rancour. It is said that he never offended anyone. Of how few men can it be recorded in words, if formal, yet of obvious sincerity, that "his memory is dear and blessed." Such a tribute lifts Palmieri from

* For instance, of 27 shocks felt in Comrie during the first half of 1841, only two were registered by the instruments.

the crowd of ordinary men into the small band of which Faraday and Darwin are honoured members*.

87. Electro-Magnetic Seismograph. The name of the instrument is of some interest. Palmieri's use of the word *seismograph* (1859) is the earliest with which I am acquainted. Since his time, the meaning of the word has changed. The instrument belongs more nearly to the class now called seismoscopes, the name seismograph being reserved for one that registers the movements of the ground with exactness and in detail.

The seismograph consists of two parts, one for the vertical, the other for the horizontal, movements. The former consists of a vertical helix of brass wire. At its upper end, it is suspended from an upright metal pillar; to the lower end is attached a cone of copper, tipped with platinum. Just below this platinum point is fixed an iron basin containing mercury; and the lengths of the pillar and helix are so adjusted that, at all temperatures, the minute distance that separates the platinum point from the mercury remains unchanged. The pillar is connected with one pole of a battery, and the mercury basin with the other. Any slight vertical movement makes the platinum point dip into the mercury, thus completing the circuit. Then two electro-magnets in the circuit attract their armatures. The movement of one armature stops the running of a clock, thus giving the initial epoch to the nearest half-second. That of the other releases the pendulum of a second clock, and this drives a band of paper over a drum at the rate of three metres an hour. At the same time, it presses a pencil against the paper, the length of the mark so made giving the duration of the shock.

For recording horizontal movements, there are four glass tubes. In each, the central part is horizontal, and the ends are bent upwards, one vertical arm being shorter than the other and having a diameter at least twice as great. The four tubes are arranged in the cardinal directions, and are partly filled with mercury. A platinum wire, connected with one pole of the battery, dips into the mercury of the wider branch; and a second wire,

* Napoli, *Rend.* vol. 2, 1896, pp. 236–238; Napoli, *Ist. Incorag. Atti*, vol. 9, 1896, pp. 12–14; *Ital. Soc. Sism. Boll.* vol. 2, 1896, pp. 157–159; *Vesuv. Oss. Ann.* vol. 1, 1859, pp. 9–10; vol. 1 (new series), 1874, pp. 97–98.

connected with the other pole, ends just above the mercury in the narrower branch. A horizontal movement will cause the mercury in one or more of the tubes to oscillate, and more sensibly in the narrower branch of each. Contact being made between the mercury and wire in this branch, the movement is recorded as before.

To determine the extent of the oscillations, an iron float rests on the surface of the mercury in the narrow branch. A fine silk thread passes from it, over an ivory pulley, to a counterpoise. The iron float thus remains at the highest point to which it is raised by the mercury, and the displacement is measured by the rotation of a long light needle attached to the axis of the pulley*.

Palmieri's seismograph thus shows the initial time of an earthquake, its duration, the existence of vertical or horizontal motion, and, very roughly, the extent and direction of the latter. That he was justified in his high opinion of its sensitiveness is clear from a comparison with a more modern instrument. During 9½ years (1875–1885), the Palmieri seismograph at Tokyo recorded 564 earthquakes. During an equal interval (1885–1894), the Gray-Milne seismograph, which replaced it, registered 842 (art. 200).

88. Records of Vesuvian Earthquakes. From 1862 onwards, Palmieri wrote many brief notes on the records given by his seismographs. The following are the principal results. (i) The shocks recorded often preceded, by a few seconds, the emission of a cloud of ashes from the crater. (ii) In the eruption that began on 8 Dec. 1861, there were two clusters of shocks, one before and with the beginning of the eruption, the other marking its end. (iii) When Vesuvius was at rest and small shocks were frequently registered at the observatory, Palmieri inferred that either a new eruption was at hand, or a distant earthquake, or an eruption of Etna or Santorin. He claimed that his seismograph would record violent earthquakes in Italy or the entire basin of the Mediter-

* *Vesuv. Oss. Ann.* vol. 1, 1859, pp. 20–24; vol. 4, 1870, pp. 41–43. Accounts in English will be found in Mallet's translation of Palmieri's memoir on the eruption of Vesuvius in 1872, pp. 141–145; J. Logan Lobley, *Mount Vesuvius*, 1889, pp. 380–385; *Smithsonian Rep.* 1870, pp. 425–428. Reference should be made to an instrument designed by Mallet (*Irish Ac. Trans.* vol. 21, 1848, pp. 107–113).

ranean; and, in several of his notes, suggested that these great earthquakes were heralded by frequent tremors recorded as much as four to six days before at the observatory. As he notes, however, that the tremors were more pronounced on Vesuvius than in Naples, and that the destructive Cosenza earthquake of 5 Oct. 1870 passed unrecorded on Vesuvius though actually felt beyond in the city of Naples, it would seem that the tremors were of local origin and were unconnected with the earthquakes*.

TIMOTEO BERTELLI AND THE FOUNDING OF MICROSEISMOLOGY

89. Not as the founder of microseismology—a branch of science lying for the most part outside the range of this book—but rather for his influence on the study of earthquakes in Italy, Timoteo Bertelli (1826–1905) and the observations that he made with such endless patience require a brief notice in these pages. He was born at Bologna on 6 Oct. 1826, his father, Francesco Bertelli, being professor of astronomy in the university of that city. In 1844, he was admitted into the Barnabite order and lectured on mathematics and physics in their houses at Moncalieri, Naples, etc. In 1868, he was transferred to the Collegio della Querce near Florence, where he remained but for one short interval for the rest of his days.

90. It was in Florence that he found time for studying the small spontaneous movements of the pendulum. Such movements had been observed for more than two centuries, and we are indebted to Bertelli for tracing the Italian records since the year 1643†.

He spent three years, from 1869 to 1872, in making a series of simple experiments, in testing the effects of changes in the place of observation or in the length of the suspending wire. At first,

* *Deutsch. Geol. Ges. Ztschr.* vol. 5, 1853, pp. 21–74; Napoli, *Acc. Atti*, vol. 1, 1888, no. 4, 28 pp.; Napoli, *Rend.* vol. 1, 1862, p. 144; vol. 3, 1864, pp. 35–36; vol. 5, 1866, pp. 102–103; vol. 6, 1867, pp. 130–131; vol. 8, 1869, pp. 98, 146–147, 179; vol. 9, 1870, pp. 127–128, 176; vol. 10, 1871, pp. 16–17, 124; vol. 14, 1875, pp. 215–216; vol. 15, 1876, p. 95; vol. 20, 1881, pp. 82–89; vol. 27, 1888, pp. 454–456; Paris, *Ac. Sci. C.R.* vol. 54, 1862, pp. 608–611; *Vesuv. Oss. Ann.* vol. 3, 1865, pp. 35–42; vol. 4, 1870, pp. 39–40.

† *Bull. Bibl. Sci. Matem. Fis.* vol. 6, 1873.

the pendulum was a ball of lead suspended from a stout arm driven into the wall of the college. Projecting downwards from the ball was a short rod ending in a point, the movements of which were observed with a lens. With this simple instrument, he made many thousands of observations. It was replaced, in June 1872, by an improved form of *tromometer*, erected on an isolated pillar in the basement of the college.

Four causes had been suggested for the movements, namely, the action of the wind, convection currents in the case surrounding the instrument, vibrations produced by passing vehicles, the expansion and contraction of the building under changes of temperature. Bertelli considered fully all such disturbing causes, and it was not until the end of November 1872 that he was able to state the following definite conclusions: (i) the microseismic movements of an isolated pendulum often occur contemporaneously with distant earthquakes; (ii) others occur during continued barometric depressions; and (iii) the movements have a maximum in winter and a minimum in summer*.

These results were disputed by Pietro Monte of Livorno, and a controversy followed into which it is now needless to enter†, needless because Bertelli's observations were repeated and confirmed by others, for instance, by Antonio Malvasia at Bologna, Ignazio Galli at Velletri, and, above all, by Michele Stefano De Rossi (1834–98) at Rocca di Papa and Rome‡.

91. In 1895, Bertelli was called to Rome as director of the Vatican observatory, but ill-health prevented him from entering into office, and, in 1898, he returned to Florence, the tromometer readings being continued by his colleague Camillo Melzi. His last years were spent in historical studies on the invention of the

* For Bertelli's principal memoirs on microseismic motions, see: Roma, *N. Lincei Atti*, vol. 27, 1874, pp. 194–208, 429–464; vol. 28, 1875, pp 334–356; vol. 29, 1876, pp. 83–110, 255–297; vol. 31, 1878, pp. 193–241.

† A summary of the controversy is given by M. S. De Rossi in *Bull. Vulc. Ital.* vol. 1, 1874, pp. 122–127; vol. 2, 1875, pp. 99–103, 107–111.

‡ Besides these early observations in Italy, reference may be made to others by G. H. Darwin (1845–1912) and H. Darwin (1851–) in England (*Brit. Ass. Rep.* 1881, pp. 93–126; 1882, pp. 95–119), and by J. Milne (1850–1913) in Japan (*Japan Seis. Soc. Trans.* vol. 11, 1887, pp. 1–78; vol. 13, 1890, pp. 7–19).

compass and on the discovery by Columbus of the magnetic declination and its variations in space. He died on 6 Feb. 1905.

Gentle and courteous in manner and careless of his own credit, Bertelli influenced a wide circle by his character and enthusiasm. In 1874, daily tromometric observations were made at five stations in Italy, ten years later at thirty. The normal tromometer is now superseded by recording instruments, but none the less must Bertelli be regarded as the founder of microseismology, of what De Rossi called the new Italian science of endogenous meteorology. As we look back on his life and work, on the thousands and thousands of measurements made with an accuracy that was part of his religion, we see in his long-continued labour one more example of the fine patience that ennobles science*.

CONTEMPORARIES OF BERTELLI

92. In addition to Bertelli, there were at this time many students of other sciences who gave some attention to earthquakes. Of two in particular—Antonio Stoppani (1824–91) and Orazio Silvestri (1835–90)—the influence extended beyond their actual work in seismology. Stoppani was professor of geology in the Higher Technical Institute of Milan. His elaborate studies on the geology of Lombardy and his volume on the Neozoic era made him widely known. His chief contribution to seismology is the chapter on earthquakes in his treatise on geology†. In this chapter, published in 1871, he divides earthquakes into three classes. (i) Volcanic earthquakes, such as those felt on the flanks of Etna, occur in volcanic districts, within, or not far beyond, the area of the volcano, and in close connexion, as regards time, with an eruption. (ii) Perimetric earthquakes, of which the Calabrian earthquakes are significant types, are confined to regions surrounding volcanoes, they shake wide areas, and, as a rule, are independent of eruptions. (iii) Telluric earthquakes, like the Lisbon earthquake of 1755, visit regions far removed from any volcano, they sometimes disturb enormous areas, and then are only repeated in the same place at long intervals of

* *Ital. Soc. Sism. Boll.* vol. 10, 1904, pp. 179–196; *Riv. Geogr. Ital.* vol. 12, 1905, 21 pp.

† *Corso di Geologia,* vol. 1, 1871, pp. 435–458.—*Geol. Soc. Quart. Jl.* vol. 47, 1891, pp. 50–51 (*Proc.*); Napoli, *Rend.* vol. 30, 1891, pp. 13–15.

time. Silvestri, by training a chemist but by inclination the student of Etna and historian of the eruptions of 1865–86, became ultimately professor of geology and vulcanology in the university of Catania. As practically founder, and for twelve years director, of the observatory near the summit of the volcano, as organiser also of the system of secondary observatories in connexion with his own as centre, Silvestri gave a distinct impulse to the study of Sicilian earthquakes*.

Besides Stoppani and Silvestri, many others might be mentioned as contributors to the rise of Italian seismology. Three of them at least require notice in this section—Filippo Cecchi (1822–87), Alessandro Serpieri (1823–85) and Francisco Denza (1834–94).

93. Outside Italy, Filippo Cecchi is best known as the designer of an electric seismograph, the objects of which were to give warning of an earthquake and to record on smoked paper its time of occurrence, its direction and intensity. He also invented a seismic adviser or seismoscope, for marking the occurrence of an earthquake, and wrote many brief notes on special earthquakes. It is perhaps as much by his influence on other workers as by his direct contributions that Cecchi is esteemed so highly in his own country†.

94. Though most of his time was given to astronomical subjects, Alessandro Serpieri, director of the meteorological observatory of Urbino, wrote several memoirs on the Urbino earthquake of 12 Mar. 1873, the Rimini earthquake of 18 Mar. 1875 and the Ischian earthquake of 28 July 1883, in the study of which he relied greatly, perhaps too greatly, on observations of the time and direction of the shock‡.

* *Rev. Sci.* vol. 46, 1890, pp. 649–651.

† *Moncalieri Oss. Boll.* vol. 7, 1887, pp. 81–82. For a description of Cecchi's seismograph, see Roma, *N. Lincei Atti*, vol. 29, 1876, pp. 421–428; also (in English) *Japan Seis. Sci. Trans.* vol. 8, 1876, pp. 90–94. An account of the adviser is given by De Rossi in *Bull. Vulc. Ital.* vol. 10, 1883, pp. 123–124.

‡ *Moncalieri Oss. Boll.* vol. 5, 1885, p. 116. Serpieri's seismological papers were collected and edited after his death by G. Giavannozzi, director of the Osservatorio Ximeniano at Florence, under the title *Scritti di Sismologia*, the first part (219 pp.) appearing in 1888 and the second (232 pp.) in 1889.

95. It is mainly by his services to astronomy that Francesco Denza will be held, and honourably held, in memory. Having entered the Barnabite order in 1850, he was trained under Angelo Secchi (1818–78) at Rome. First among his important works was the foundation of the meteorological observatory of Moncalieri, near Turin, in 1859; while the crowning achievement of his life was the successful restoration of the Vatican observatory. The monthly *Bollettino* of the Moncalieri observatory, first published in 1866, gave an impulse to the study of meteorology in Italy. Indirectly, it also affected that of seismology. From 1874 to 1879, Denza sent notices to De Rossi's *Bullettino del Vulcanismo Italiano* (art. 99), after which he wrote about a dozen papers in his own *Bollettino,* in which various earthquakes were briefly described. In the pages of this journal, he also inserted articles by Galli, De Rossi, Silvestri, Stoppani and other writers. Few men can have left behind them such a record of energetic labour and of unfailing kindness to those who sought his help*.

MICHELE STEFANO DE ROSSI and the *BULLETTINO DEL VULCANISMO ITALIANO*

96. To few men, perhaps to no man, is Italian seismology so deeply indebted as to Michele Stefano De Rossi (1834–98). Others since his death may have contributed more notably to the science. It was De Rossi's good fortune that he came at a time when seismology was in the making and that, by his energy and enthusiasm, he was able to contribute to its advance.

Born in Rome on 30 Oct. 1834 and educated in the university of that city, De Rossi devoted himself at first to the topography of the catacombs†. From the antiquities of the Roman Campagna, he passed on to its geology and, rather late in life, to its earthquakes. His interest in them dates from the year 1868, when he felt an earthquake that disturbed Altorf on the north and Siena on the south. From this shock, or rather from the neglect of it, he realised that Italy possessed no organisation for

* *Meteor. Soc. Quart. Jl.* vol. 21, 1895, pp. 75–76; *The Observatory,* London, vol. 18, 1895, p. 47; Symons, *Meteorol. Mag.* vol. 29, 1895, p. 177.

† His first paper on the antiquities of the Roman Campagna was published in 1866, the last in 1886.

the study of earthquakes, and, in a letter to a Genovese paper, he urged that the want should be supplied. Nearly five years passed before a remarkable group of earthquakes in the Alban Mountains near Rome on 19 Jan. 1873 led him "almost against his will" to the detailed investigation of earthquakes. The memoir in which he described these earthquakes was followed in the same year by another on the earthquakes of 7 Feb.–30 Apr. 1873, and, through them, De Rossi was recognised as the leading student of Italian earthquakes. Two other early memoirs, both published in 1875, are worthy of mention, one dealing with three important Italian earthquakes of 1874 with special reference to fractures of the ground, the other with the earthquakes of Romagna from Sep. 1874 to May 1875*.

97. Attracted by the researches of Timoteo Bertelli, De Rossi's early observations were made on similar lines, partly at his house in Rome and in the catacombs, but chiefly in his villa at Rocca di Papa, soon to become almost as widely known as Palmieri's observatory on Vesuvius or as Milne's in the Isle of Wight. Here, in a cave hollowed out of the side of an extinct crater of the Alban Mountains, De Rossi erected his instruments, repeated and extended Bertelli's observations and became firmly convinced of the reality of the microseismic motions. Here, also, he placed several new instruments that he devised—the auto-seismograph, the protoseismograph and the microseismograph† —and, here and elsewhere, he applied the microphone to the study of subterranean sounds‡.

In 1874, De Rossi began the issue of the *Bullettino del Vulcanismo Italiano*, the first journal devoted to the study of volcanoes and earthquakes (art. 99). This was followed, a few years later, by a textbook on "endogenous meteorology." This name he preferred to the older one of seismology, though afterwards he adopted the more general term geodynamics, now so widely used. As he claims that "the science of endogenous meteorology is

* Roma, *N. Lincei Atti*, vol. 26, 1873, pp. 136–179, 262–299; vol. 28, 1875, pp. 14–87, 308–333.

† Described in *Bull. Vulc. Ital.* vol. 1, 1874, pp. 141–148; vol. 4, 1877, pp. 34–39; vol. 9, 1882, pp. 143–145.

‡ *Bull. Vulc. Ital.* vol. 5, 1878, pp. 53–62, 99–120; vol. 9, 1882, pp. 143–145; vol. 10, 1883, pp. 105–106, 139–141.

the result of recent Italian studies," it is only natural that *La Meteorologia Endogena** should consist almost entirely of summaries of his own work and of that of Bertelli, Monte, etc. To the achievements of foreign students little attention is paid, and thus its two volumes can hardly be regarded as a textbook of seismology in general.

98. It was through the *Bullettino* that De Rossi's conception of a national geodynamic service was realised in part. He was aided by the interest aroused by several strong earthquakes, by none more than those which damaged Casamicciola in 1881 and ruined it in 1883. A result of the former shock was the foundation, with official approval, of a geodynamic section of the R. Comitato geologico, known as the *Archivio centrale geodinamico,* placed under De Rossi's charge. Rooms were allotted to him provisionally in the new building erected for the Geological Commission in 1883, and here his instruments were established for the next seven years.

It is difficult to understand De Rossi's exact position at this time. He held a government appointment and occupied an office in a government building, but he seems to have had no staff and no organisation beyond that which he had himself created. His appointment was perhaps intended to be of a less permanent character than he at the time imagined; for, soon after the Ischian earthquake of 1883, a royal commission was nominated to found a geodynamic service for the whole of Italy (art. 112). Of this commission, De Rossi was a member; and, if an independent service could have been created, he might reasonably have expected the first directorship. Unfortunately, it was necessary, from motives of economy, to link the new service with the already existing Meteorological Office, and thus the opportunity for extending the work he had carried out with so much zeal and success passed from his to other hands.

To De Rossi, this decision must have been one of life's great

* Vol. 1, 1879, pp. xv + 359; vol. 2, 1882, pp. xii + 437. The work forms one of the well-known *International Scientific Series*, but it was never translated into English. By the time that the second volume appeared, the English publishers had probably arranged with Milne for his volume on *Earthquakes and other Earth Movements*, published in 1886, though, as the preface shows, written in 1883.

disappointments. It was but a poor compensation to be offered the charge of a new observatory at Rocca di Papa, of an unit instead of the entire service. Such as it was, however, De Rossi accepted it, and in 1890, as soon as the building was finished, his instruments were transferred to their last home. But his work by this time was nearly done. After 1892, illness and domestic sorrows lessened his abounding energy, and six years later, on 23 Oct. 1898, he died at Rocca di Papa*.

The chief work of De Rossi's life was no doubt the organisation that he had hoped to transform into a national service. To have founded it at all is evidence of the place that he held in the esteem of his countrymen, of his intense energy and self-reliance. To have maintained and expanded it required unfailing tact and courtesy. Some of the correspondents who helped him in the last volume of the *Bullettino* (1897) were with him at the start—men such as Bertelli, Denza, Galli, Palmieri and Silvestri —and we could hardly desire a clearer token than this of the value set on De Rossi's work or of the strength of his influence.

99. Bullettino del Vulcanismo Italiano. If De Rossi had done nothing else, he would have deserved our gratitude as the founder of the first journal devoted to "the science of the endogenous forces of the earth." His chief object in starting it was to encourage the spread of regular observations such as were then being made by Palmieri on Vesuvius, Bertelli in Florence and Denza in Moncalieri. At the time, it was a remarkable venture, and it is not surprising that the cost of its production should have fallen in part on its enthusiastic editor.

The first volume was published in 1874, and one in each succeeding year until the fourteenth in 1887. Vols. 15 and 16, for 1888 and 1889, appeared as one, and vol. 17 in 1890. Then came a long pause, and it was not until 1897 that the series was concluded by the issue of vols. 18–20 as little more than a pamphlet of eighty pages†.

* *Ital. Soc. Sism. Boll.* vol. 4, 1898, pp. 105–106; vol. 15, 1911, pp. 111–124; *Bull. Paletn. Ital.* vol. 3, 1897, pp. 310–312; Roma, *N. Lincei Atti*, vol. 52, 1899, pp. 37–57, 74–97.

† In the first eight volumes (1874–81), the sub-title of the journal is "Periodico geologico ed archeologico per l' osservazione e la storia dei fenomeni endogeni nel suolo d' Italia," in the next six volumes (1882–87),

The structure of the journal, as outlined at the start, continued throughout with little change. As a rule, each part opened with one or more articles, followed by several pages of bibliography, and these by letters from the observers who agreed to cooperate with De Rossi. Then came an important section in which the observations made at the various stations were tabulated, and, lastly, a part, always brief and often absent, on historical notices of former endogenous phenomena in Italy. Roughly speaking, out of every ten pages, four were occupied by articles, one by notices of memoirs, one by correspondence, and four by tabular summaries of the observations.

Of the total number of articles, four-fifths are concerned with earthquakes or the minute movements of the ground, the rest with volcanic eruptions, etc. Many of the former are still of interest, especially the following written mainly or entirely by De Rossi—a practical guide for seismic observations, the microphone in endogenous meteorology, the programme of the central geodynamic observatory and archives, and, above all, the summary of memoirs on the Ischian earthquake of 1883*. In an early journal on earthquakes, it is only natural that many new instruments should be described. Accounts are given in various volumes of ten seismic advisers or seismic spies—instruments that are designed to note the occurrence and time of a shock, and that would now be called seismoscopes—nine seismographs of various patterns, four tromometers, and four seismic microphones. If few of these instruments are now in use, they none the less served a purpose in the years that preceded the invention of accurate seismographs in Japan.

During the first year (1874), the number of observers who cooperated with De Rossi was 23, including such well-known men as Bertelli, Denza, Galli, Issel, Malvasia, Palmieri, Serpieri and

concurrently with De Rossi's temporary appointment at Rome, it was changed to "Periodico dell' osservatorio ed archivio centrale dei fenomeni endogeni in Italia presso il R. Comitato Geologico"; in the remaining volumes (1888–90, 1897), the words "e di Geodinamico Generale" were added instead of the above to the main title, the name of the editor being followed by the words "founder and director of the Osservatorio ed Archivio Geodinamico in Roma."

* Vol. 4, 1877, pp. 5–42; vol. 5, 1878, pp. 99–120; vol. 10, 1883, pp. 3–144; vol. 11, 1884, pp. 65–172.

Silvestri. By the fourth year (1877), it had risen to 105, Egidi, Grablovitz, Mocenigo, Monte, Stoppani and Taramelli being among the recruits. In the fourteenth year (1887), the list contains 133 observers, of whom 66 possessed instruments for direct observation, 119 seismoscopes, and 48 apparatus with continuous registration. Concurrently with the increase in the number of corresponding observers, there was a growth in the number of those who took daily readings of the tromometer. At the end of 1874, there were five such observers—at Rome, Rocca di Papa, Florence, Livorno and Bologna. In 1877, the number had risen to 10; in 1880, to 14; in 1883, to 20, and, at the beginning of 1885, to 30, after which it slightly declined.

De Rossi's method of presenting his results varied. From Dec. 1872 to July 1887, we have a list of all known Italian earthquakes, and, from Dec. 1874 to Nov. 1889, measurements of the small displacements of the pendulum at different observatories. How valuable the tables are will be evident from the statement that, during the first eight years (1873–80), the total number of earthquakes recorded fell just short of six thousand.

During the most vigorous period of the *Bullettino*, that is, in the first fourteen years, the total number of pages issued was more than 2400. In the period of decline, covering that of the last six volumes, the number did not rise beyond 320. About two-thirds of the articles, occupying 85 per cent. of the space, were either written or compiled by De Rossi. The rest were contributed by 23 other authors.

The decline of the *Bullettino* was due to three causes. (i) Being mainly the work of one man, the supply of material shrank with his failing health and energy. (ii) Towards the close, the journal appeared at long and fitful intervals, and Denza's monthly *Bollettino* of the Moncalieri observatory offered a more prompt mode of publication for short papers. But (iii) the most important cause was the competition from 1889 onwards of the *Bullettino* of the Central Office of Meteorology and Geodynamics. Supported by government funds, this produced more rapidly and effectively the issue of notices for which De Rossi's *Bullettino* was mainly designed. At the same time, the *Annali* of the Central Office found room for memoirs that several volumes of the

Bullettino would have failed to hold. If, however, the latter was finally superseded by its own more vigorous offspring, it must not be forgotten that it was De Rossi's journal that helped to create the public interest that made the national service possible.

100. Scales of Seismic Intensity. Long before De Rossi's time, scales of seismic intensity were proposed, either to express the relative intensity of the different members of a series of shocks, as in the Pignataro scale of 1783 (art. 29), or to trace the variation of intensity of a single earthquake throughout its disturbed area, as in the Egen scale of 1828 (art. 42). Mallet had also devised two scales, that of 1858 with the former object, and that of 1862 with the latter (art. 73). To all the scales before 1874, however, one common feature belonged—not one of them was used by anyone but its author. It was reserved for De Rossi to introduce a scale that met with general approval.

In the first volume of his *Bullettino*, De Rossi gave a list of the Italian earthquakes of the year 1873, and, in describing them, he tried to give a definite meaning to every adjective used. It is evident that his thought was to frame a scale for comparing the intensities of different earthquakes. The earliest form of the scale is here given. In subsequent years, De Rossi modified the degrees 3–8 of his scale by the addition of new tests, but it is unnecessary to trace these changes from year to year*.

DE ROSSI SCALE OF INTENSITY

1. Very slight shock, recorded by seismographs or by one seismologist.
2. Weak shock, felt by more than one.
3. Slight shock, felt by many.
4. Sensible shock, accompanied by shaking of fastenings, chandelier-prisms and furniture.
5. Moderate shock, felt generally by very many persons.
6. Rather strong shock, ringing of bells, oscillation of lamps, stopping of clocks.

* *Bull. Vulc. Ital.* vol. 1, 1874, p. i; vol. 2, 1875, p. iii; vol. 3, 1876, p. i; vol. 4, 1877, pp. 39–40; vol. 5, 1878, p. 46.

7. Strong shock, fall of plaster, ringing of church bells, noise.
8. Very strong shock, fall of chimneys and cracks in buildings.
9. Ruinous shock, total or partial fall of some buildings.
10. Disastrous shock, great ruins and loss of life.

101. In 1878, the Helvetic Society of Natural Sciences appointed a committee for the study of earthquakes in Switzerland (arts. 138–141). From the first, a prominent part in its work was taken by François Alphonse Forel (1841–1912) of Lausanne. Unaware of the existence of the De Rossi scale, Forel proposed a similar one. This also consists of ten degrees, which, though depending partly on personal impressions, are more definite than those of the De Rossi scale*. Soon after its publication, Forel and De Rossi each became acquainted with the scale suggested by the other. Both contained points worth preserving and, on the invitation of De Rossi, the two seismologists met in order to agree on a single scale for the earthquakes that occurred on both sides of the Alps. Their success was probably greater than they anticipated for the resulting scale has been more widely used than any other. Though it is not without defects, it would be difficult, I think, to over-estimate the value of the results that have followed from its use. It should be noticed that the scale given below is translated from the Swiss version†. In the *Bullettino del Vulcanismo Italiano*, De Rossi states that the changes were made in agreement with the Swiss Seismological Commission. The two versions, Italian and Swiss, are given in parallel columns; and it is curious to note that, while there is really no essential difference between them, there are several variations in points of detail. On the whole, the Italian version approximates more closely to the last form of the De Rossi scale (1878) than to the Swiss version of the combined scales. The latter no doubt owes its wider circulation to its publication in the French language and in a more accessible journal; but, even if it had been otherwise, there can, I think, be little doubt that it is the superior form.

* *Arch. Sci. Phys. Nat.* vol. 6, 1881, pp. 465–466.

† *Bull. Vulc. Ital.* vol. 10, 1883, pp. 67–68; *Arch. Sci. Phys. Nat.* vol. 11, 1884, pp. 148–149. The Italian version is repeated in vol. 12, 1885, p. 8.

ROSSI-FOREL SCALE OF INTENSITY

1. Recorded by a single seismograph, or by some seismographs of the same pattern, but not by several seismographs of different kinds; the shock felt by an experienced observer.

2. Recorded by seismographs of different kinds; felt by a small number of persons at rest.

3. Felt by several persons at rest; strong enough for the duration or direction to be appreciated.

4. Felt by several persons in motion; disturbance of movable objects, doors, windows; creaking of floors.

5. Felt generally by every one; disturbance of furniture and beds; ringing of some bells.

6. General awakening of those asleep; general ringing of bells; oscillation of chandeliers, stopping of clocks; visible disturbance of trees and shrubs; some startled persons leave their dwellings.

7. Overthrow of movable objects, fall of plaster, ringing of church bells, general panic, without damage to buildings.

8. Fall of chimneys, cracks in the walls of buildings.

9. Partial or total destruction of some buildings.

10. Great disasters, ruins, disturbance of strata, fissures in the earth's crust, rock-falls from mountains.

With various modifications made to suit local conditions, the Rossi-Forel scale has been used in the investigation of earthquakes in many countries in Europe and America, though, in Italy, it has been replaced by the Mercalli scale (art. 104). On it was founded the first absolute or dynamical scale. In 1888, E. S. Holden (1846–1914; art. 150) estimated the maximum accelerations corresponding to degrees 1 to 9 of the Rossi-Forel scale as, respectively, 20, 40, 60, 80, 110, 150, 300, 500 and 1200 mm. per sec. per sec.*

102. It is not easy to assess the value of De Rossi's contributions to seismology. Apparently, his life ended in failure. His early studies on earthquakes, useful as they were, have left no permanent mark on the science. They are seldom referred to now and find no place in modern textbooks. Of the instruments that

* *Amer. Jl. Sc.* vol. 35, 1888, pp. 428–429.

he devised, probably not one is in use at the present day. They were not based on those sound mechanical principles that would have made them a lasting possession. After 1889, the usefulness of the *Bullettino* ended, and, a little later, the last parts appeared. The great aim of his life, the conduct of a national geodynamic service, was carried out by another.

What have we left? Chiefly, it seems to me, the influence of the energy and enthusiasm of an earnest and unresting worker, the atmosphere diffused by the man who lives one life, not two or three. It is this that is so incalculable. But, if we think how few of his countrymen were interested in earthquakes in 1874, how many were enlisted under his guidance and, twenty years later, were spending their lives in the study to which he devoted his, we shall realise that seismology in Italy would be in quite a different position now if De Rossi had never lived.

GIUSEPPE MERCALLI AND THE REGIONAL DISTRIBUTION OF EARTHQUAKES

103. The importance of regional memoirs on earthquakes had long been recognised, by Élie Bertrand in 1757 and by Alexis Perrey during the years 1845–66 (arts. 8, 9, 52). Their work, however, consisted mainly in cataloguing the earthquakes of limited districts. It was reserved for Giuseppe Mercalli (1850–1914) to study the distribution of earthquakes in another and original way. In addition to two important memoirs on this subject, he investigated several great earthquakes, and studied the phenomena of the Italian volcanoes and especially of Vesuvius. And all this was done, it should be remembered, in the leisure hours of a busy teacher's life.

Giuseppe Mercalli was born at Milan on 21 May 1850 and was educated at the normal school of the Higher Technical Institute in that city. Graduating as professor of natural science in 1874, he taught for some years in the Seminario of Monza and in private schools of Milan. In 1889, he was elected professor in the Liceo Campanella of Reggio Calabria; in 1891, at the university of Catania; and, in 1893, in the R. Liceo Vittorio Emanuele of Naples. Many years later, in 1911, he succeeded to the directorship of the Vesuvian observatory. Thus, for a quarter of a century,

Mercalli lived and worked close to the principal seismic and volcanic districts of Italy.

His intimate knowledge of these districts dates, however, from 1878, when he visited all the Italian volcanoes in order to obtain materials for his work *I Vulcani e Fenomeni vulcanici in Italia* (1883). From 1878 to 1891, he went many times to the Aeolian Islands, and studied especially the phenomena of Vulcano and Stromboli. He also devoted some time to the eruptions of Etna and published a detailed account of the outburst of 1892. But to no volcano did he give such close attention as to Vesuvius. From 1893, when he came to live in Naples, he followed all its changing phases, and especially the great eruption of 1906 and its succeeding period of repose.

Mercalli's interest in the earthquakes of Italy began at about the same time as his studies of volcanoes. The second part of the volume just referred to contains a catalogue of Italian earthquakes and a series of maps in which he depicted the distribution of seismic activity in four successive periods. Both catalogue and maps have served as the foundations for similar works, and especially for that of Mario Baratta published in 1901*.

Mercalli investigated several important earthquakes, especially those of Ischia in 1883, Andalusia in 1884, the Riviera in 1887, Calabria in 1894, 1905 and 1907, and Messina in 1908. It was characteristic of him that he felt his studies to be incomplete without delving into the seismic history of the respective districts. His two great regional monographs were published in 1897, the first on the earthquakes of Liguria and Piedmont, the second on those of southern Calabria and the Messinese district.

On 19 Mar. 1914, Mercalli's useful life was abruptly ended by the accidental overthrow of a petroleum lamp. Simple in his habits, of upright character and courteous manner, always ready to give help to others, Mercalli seems to have been esteemed as much for his personal charm as for the results of his unceasing labour†.

* *I Terremoti d'Italia*, Turin, 1901, 951 pp.

† Napoli, *Rend.* vol. 53, 1914, pp. 21–25; *Ital. Soc. Sism. Ital.* vol. 17, for 1913, pp. 245–262.

104. Scales of Seismic Intensity. As an investigator of earthquakes, Mercalli soon realised the usefulness of some scale of intensity. His first scale, used in his catalogue of Italian earthquakes (1883), is an adaptation of the De Rossi scale (art. 100), and bears a close resemblance to the Saderra Masò and Rockwood scales (art. 148), all three containing only six degrees*.

A few years later (1887), while studying the Riviera earthquake, Mercalli found that the higher degrees of the Rossi-Forel scale (art. 101) were insufficient to discriminate variations in the amount of damage. His modification of that scale† is interesting, as it contains the germs of the Mercalli scale now so widely used.

This final form of the scale was proposed in 1897‡. It was designed for use in a country visited by strong earthquakes and has been adopted as the standard of the Central Office of Meteorology and Geodynamics at Rome. In the following translation, the corresponding degrees of the Rossi-Forel scale are added in brackets:

MERCALLI SCALE OF INTENSITY

1. Instrumental shock, that is, noted by seismic instruments only (1).

2. Very slight, felt only by a few persons in conditions of perfect quiet, especially on the upper floors of houses, or by many sensitive and nervous persons (2).

3. Slight, felt by several persons, but by few relatively to the number of inhabitants in a given place; said by them to have been *hardly felt*, without causing any alarm, and in general without their recognising that it was an earthquake until it was known that others had felt it (3).

4. Sensible or moderate, not felt generally, but felt by many persons indoors, though by few on the ground-floor, without causing any alarm, but with shaking of fastenings, crystals, creaking of floors, and slight oscillation of suspended objects (4, 5).

* *Vulcani e fenomeni vulcanici in Italia*, pp. 217–218.

† Roma, *Uff. Centr. Meteorol. Ann.* vol. 8, pt. 4, 1888, pp. 60–61.

‡ *I Terremoti della Liguria e del Piemonte*, Naples, 1897, pp. 19–20; *Ital. Soc. Sism. Boll.* vol. 8, 1902, pp. 184–191. The Mercalli scale has been modified by A. Sieberg (*Beitr. Geophys.* vol. 11, 1912, pp. 227–239).

5. Rather strong, felt generally indoors, but by few outside, with waking of those asleep, with alarm of some persons, rattling of doors, ringing of bells, rather large oscillation of suspended objects, stopping of clocks (6).

6. Strong, felt by everyone indoors, and by many with alarm and flight into the open air; fall of objects in houses, fall of plaster, with some cracks in badly-built houses (7).

7. Very strong, felt with general alarm and flight from houses, sensible out-of-doors; ringing of church bells, fall of chimney-pots and tiles; cracks in numerous buildings, but generally slight (8).

8. Ruinous, felt with great alarm, partial ruin of some houses, and frequent and considerable cracks in others; without loss of life, or only with a few isolated cases of personal injury (9).

9. Disastrous, with complete or nearly complete ruin of some houses and serious cracks in many others, so as to render them uninhabitable; a few lives lost in different parts of populous places (10).

10. Very disastrous, with ruin of many buildings and great loss of life, cracks in the ground, landslips from mountains, etc. (10).

In 1904, A. Cancani (1856–1904) suggested an absolute scale of intensity which bears the same relation to the Mercalli scale as the Holden scale (art. 150) does to the Rossi-Forel scale. He added two degrees, however, to represent the highest of all intensities. The maximum accelerations (in mm. per sec. per sec.) corresponding to the different degrees are as follows: 2·5, 5, 10, 25, 50, 100, 250, 500, 1000, 2500, 5000 and 10,000*.

105. Catalogue of Italian Earthquakes. In 1883, at the age of thirty-three, Mercalli published his remarkable work *I Vulcani e Fenomeni vulcanici in Italia*†. With the first part (pp. 1–215) on the volcanoes and allied phenomena we have here no concern. The second part (pp. 216–367) is devoted to the earthquakes and their distribution in time and space. The

* *Compt. Rend. des Séances de la 2me Conf. Sismol. Intern.* 1904, pp. 281–283 (published as a supplement of *Beitr. Geophys.*).

† Part 3 of *Geologia d'Italia*, by G. Negri, A. Stoppani and G. Mercalli, 1883, 376 pp.

catalogue of Italian earthquakes (pp. 219–332, 360–365) is one of the earliest and most successful attempts to chronicle the earthquakes of the country as a whole, containing, as it does, notices of more than five thousand earthquakes from 1450 B.C. to A.D. 1881*.

As in all European catalogues, the number of entries increases rapidly towards its close. Two-thirds of the total number, indeed, occurred during the last fifty years. Of disastrous and ruinous earthquakes only, there were 277 in the last 480 years, namely, 31 in the fifteenth century, 28 in the sixteenth, 41 in the seventeenth, 88 in the eighteenth, and 89 in the nineteenth up to 1880. The catalogue contains notices of about 210 shocks in 1869, 350 in 1870 and 330 in 1871; and thus, as the true number must be far greater, Mercalli concluded that a sensible earthquake shakes the ground of Italy almost every day.

In the disastrous and ruinous earthquakes Mercalli failed to detect any tendency to periodicity. But he noticed a marked tendency to clustering in certain groups of years. For instance, of 40 such earthquakes in the seventeenth century, 30 occurred within 21 years; of 88 in the eighteenth century, 67 in 40 years; and of 89 in the nineteenth century, 85 in 42 years. Or, stating the case in another way, the number of earthquakes in the years of condensation was nine times as great as during an equal number of the remaining years.

In Mercalli's discussion of his catalogue, by far the most interesting and original feature is his method of representing the distribution of seismic activity in Italy (pp. 351–355). He divided the country into thirty seismic districts, in each of which earthquakes are repeated with more or less similar characters. Then, by different tints of shading, he represented ten degrees of an arbitrary scale, degree 1 corresponding to some strong earthquakes or to one strong earthquake with several minor shocks, and degree 10 to from 8 to 10 disastrous or ruinous shocks. In this way, he drew four maps of the country corresponding to the following periods: 1303–1499, 1502–1631, 1632–1737 and 1750–

* Mercalli's principal sources are the catalogues of Hoff, Perrey and Mallet, and the lists for certain limited regions of G. Baglivi, E. Capocci, C. Gemellaro, A. Goiran, G. Guarini, L. Pilla and A. Serpieri (arts. 85, 95).

1849. From these maps, Mercalli draws some interesting conclusions. He showed that, in any period, the seismic activity differs much in adjacent regions. It is very great in some districts far removed from volcanoes, such as the Calabrias, Romagna, etc. From one epoch to another, the seismic activity of a district varies considerably, and, in neighbouring districts, sometimes in opposite directions. Taking the country as a whole, the seismic activity does not change much from one epoch to another, though, in the last two periods, there is noticeable a slight decrease in northern Italy, and a slight increase in the south. Mercalli connected the latter with the frequent activity of Vesuvius during the last two periods.

106. Investigation of certain Earthquakes. Mercalli's first paper on seismology dealt with the earthquakes of the island of Ischia, 20 in number from July 1228 to Mar. 1881*. The severe shock of the latter year was followed by the violent earthquake that destroyed Casamicciola on 28 July 1883. This was the first earthquake that Mercalli studied in the field. His report on it is a memoir of more than fifty quarto pages†. The methods of investigation that he employed were mainly those devised by Mallet (arts. 70, 71). He found that 46 lines of direction converged towards Casamenella, intersecting within a narrow band that coincides with a radial fracture of the volcano Epomeo. Again, from five measurements of the angle of emergence, he estimated the mean depth of the focus to be 1·2 km. or about three-quarters of a mile. As all the great earthquakes of Ischia possess similar features—the same epicentre, the restricted area of disturbance, the sudden onset of the shock, etc.—he concluded that the earthquake of 1883, like its predecessors, was a true volcanic earthquake, an unsuccessful attempt to force an eruption‡.

By this memoir, Mercalli had proved himself a careful observer, and, early in 1885, he was chosen by the Minister of

* Milano, *Soc. Ital. Atti*, vol. 24, 1881, pp. 20–37.

† Milano, *Ist. Lomb. Mem.* vol. 15, 1884, pp. 99–154.

‡ Some years later, Mercalli studied a group of nine similar, but slighter, shocks felt on 15–16 Nov. 1892 in the island of Ponza, about 50 miles west of Ischia (Napoli, *Acc. Atti*, vol. 6, 1894, no. 10, 27 pp.).

Public Instruction to accompany T. Taramelli (1845–1922) in the investigation of the great Andalusian earthquake of 25 Dec. 1884. Of their valuable report*, the part dealing with the earthquake was the work of Mercalli. On the map of the earthquake, he drew two isoseismal lines, from the form of which and from 47 observations of the direction, he assigned to the epicentral area an elliptical form, 8½ miles long from east to west, and 2 or 2½ miles wide, its centre being about 24 miles north-east of Malaga. Thirteen estimates of the angle of emergence gave depths ranging from 7·7 to 22·7 km. with an average of 12·3 km. or 7¾ miles.

107. Two years later, Taramelli and Mercalli were again called on to investigate an important earthquake, that of the Riviera on 23 Feb. 1887. Of their great report†, one of the most complete presented on any earthquake, Mercalli wrote the seismic section (pp. 46–296). On his map of the earthquake, he drew five isoseismal lines, the first bounding the area in which the shock was disastrous or ruinous, an area extending along the coast for about 105 miles, but inland only from 9 to 12 miles. This form shows that the epicentre was submarine. Observations on the direction were made at 120 places, most of the lines intersecting within a narrow band parallel to the coast, the centre of which lies 15 miles south of Oneglia. Several lines of direction, however, had no connexion with this epicentre, and Mercalli traced them to a secondary epicentre lying to the south of Nice, an inference that received support from observations on the time and on the relative intensity of the two parts of which the shock consisted. The depth of the principal focus was found, from three angles of emergence, to lie between 16·8 and 18·8 km., the average being 17·5 km. or nearly 11 miles.

108. After 1892, Mercalli's investigations in the field were confined to the earthquakes of southern Calabria and Messina. The report on the destructive earthquake of 16 Nov. 1894 forms the third part of his regional memoir on these earthquakes‡. The epicentral area, bounded by the innermost isoseismal, overlaps, but does not quite coincide with, that of the great earthquake

* Roma, *R. Acc. Lincei Mem.* vol. 3, 1886, pp. 116–221.

† Roma, *Uff. Centr. Meteorol. Ann.* vol. 8, pt. 4, 1888, 298 pp.

‡ Roma, *Soc. Ital. Mem.* vol. 11, 1897, pp. 116–143.

of 5 Feb. 1783. For determining the position of the epicentre, Mercalli still relied on observations of the direction, but he made no attempt to ascertain the depth of the focus. The lines of direction (33 in number at 18 places) intersect for the most part in two places, one on the western slope of Aspromonte between S. Cristina and Delianova, the other a few kilometres out at sea between Palmi and Capo Peloro. To the existence of the corresponding foci, he attributes the two parts of the shock that were observed over a wide area.

Of the Calabrian earthquakes of 8 Sept. 1905 and 23 Oct. 1907, we have only preliminary notes on Mercalli's observations*. In area of destruction, the former shock has been exceeded by no other Calabrian earthquake, not even by those of 27 Mar. 1638 and 5 Feb. 1783. This area is more than 60 miles long from north to south and overlaps the western, but not the eastern, coast. Mercalli considered that it contained at least two centres, to the north and south of the isthmus of Catanzaro. The earthquake of 1907 originated on the opposite coast. Its epicentral area was unusually small, including only the district round Ferruzzano, and thus, as Mercalli inferred, the depth of its focus must have been small.

109. Lastly, we have the Messina earthquake of 28 Dec. 1908, investigated by Mercalli in the following April†. On his map of the earthquake, he gave four isoseismal lines of intensities 8–11 (Mercalli scale extended by Cancani). The meizoseismal area he found to be an ellipse 18–20 km. long from north to south and about 10 km. wide, with its axis rather nearer the Calabrian, than the Sicilian, coast. This earthquake obliged Mercalli to add one more centre (the Reggio-Messina) to the eighteen seismic centres defined in his regional memoir on the Calabrian earthquakes (art. 111). To this centre, he referred, among other earthquakes, two of the strongest after-shocks of the Calabrian earthquake of 5 Feb. 1783, namely, those of 7 Feb. at 22 h., and 11 June, 1783.

* Napoli, *Acc. Ponton. Atti*, vol. 36, 1906, 9 pp.; *Ital. Soc. Sism. Boll.* vol. 13, 1908, pp. 9–15. The complete reports in the Central Office of Meteorology and Geodynamics are not yet published.

† Napoli, *Ist. Incorag. Atti*, vol. 61, 1910, pp. 249–292.

110. Regional Distribution of Earthquakes. On this subject Mercalli published two important memoirs, both in 1897, one on the earthquakes of Liguria and Piedmont, the other on those of Calabria and Messina (see also art. 105).

The first memoir*, published at his own expense, contains a catalogue of three earthquakes before the Christian era and of 1638 earthquakes from 951 to 1895, the latter including four disastrous, and 15 ruinous, earthquakes. In the present section, we are concerned with the distribution of these earthquakes in space. Mercalli defined a *seismic centre* as any cause that is capable of producing earthquakes and that acts in a determinate place, and a *seismic district* as a group of seismic centres, the boundary being governed by orographical and geological conditions. By plotting the central areas of 180 earthquakes on a map of the region, he distinguished altogether 12 seismic districts, of which five are in Liguria and seven in Piedmont. To represent the relative seismicity of these districts, he used the same method and the same scale as for the map of Italy (art. 105), the highest degree attained being the eighth (corresponding to four or five disastrous or ruinous earthquakes). Three maps are drawn for the sixteenth, seventeenth, eighteenth and nineteenth, centuries, respectively. They show how, in each district, the seismic activity varies from one epoch to another. In the sixteenth and seventeenth centuries, it was greatest in the Nice district (degree 8); in the eighteenth century, in the Piedmontese valleys (degree 4–5); and, in the nineteenth century, in western Liguria (degree 8). Thus, the maximum seismicity corresponds to the mountainous nucleus of the Maritime Alps. As a rule, the central areas of the most important earthquakes are elliptical in form, the longer axes in the more frequent transversal earthquakes being parallel to the transverse Alpine valleys; but, in the rarer longitudinal earthquakes (e.g. that of 23 Feb. 1887), to the axes of the Apennine or Alpine chains.

111. Equal in number and far more violent were the earthquakes of southern Calabria and Messina†. Mercalli's catalogue

* *I Terremoti della Liguria e del Piemonte*, 1897, 147 pp. and 3 plates.
† Roma, *Soc. Ital. Mem.* vol. 11, 1897, pp. 117–266, 2 plates.

contains records of more than 1600 earthquakes from 1169 to 1895, of which eight were very disastrous, 20 disastrous and 13 ruinous. In the whole region, he defined 18 seismic centres (to which he afterwards added the Reggio-Messina centre: art. 109). He was unable, however, to mark out the seismic districts, as the epicentral areas of so many earthquakes were unknown. He therefore drew three maps on which are shown those areas for the more important earthquakes of the sixteenth and seventeenth centuries, the eighteenth century, and the nineteenth century. These maps show at once the much higher seismicity of the western, than of the eastern, slope of the Calabrian province. In the seventeenth century, the most active centre was the isthmus of Catanzaro; in the eighteenth, the Palmi-Oppido plain; in the nineteenth century, the activities of both centres, though of a lower order, were nearly equal. Of remarkable interest is Mercalli's discussion of the greatest of the Calabrian series, that of 1783, and of the migration of the earthquake-foci. He was, I believe, the first to distinguish between the two disastrous earthquakes of 7 Feb. at 20h. 20m. and 22h., regarded as one by the early investigators in spite of the divergence of the times and the difference of the central areas (arts. 27, 29, 32). He assigned the earthquakes of 5 and 6 Feb. to the Calabrian plain, the earlier shock of 7 Feb. to the neighbourhood of Soriano and the later to that of Messina, the earthquake of 1 Mar. to the valley of the Angitola, that of 28 Mar. to the isthmus of Catanzaro, and the almost ruinous shock of 29 July to the neighbourhood of Gerace on the eastern slope of Aspromonte. Lastly, Mercalli showed that the seismicity of the Lipari Islands, with their two active volcanoes, is far inferior to that of the neighbouring coasts of Sicily and Calabria. No ruinous earthquake is known to have occurred in the islands of Stromboli and Vulcano, and only one violent earthquake (on 16 Mar. 1892) in the extinct volcanic foci of Alicuri and Filicuri.

PIETRO TACCHINI AND THE CENTRAL OFFICE OF METEOROLOGY AND GEODYNAMICS

112. The share of Tacchini (1838–1905) in the progress of seismology was administrative rather than direct. His own contributions were limited to nine brief notes, and, in these, he appears more as the head of an office than as an investigator of earthquakes.

Pietro Tacchini was born at Modena on 21 Mar. 1838. He received his astronomical training in the observatory of Padua, and for twenty years (1859–79) he worked in the observatories of Modena and Palermo. In the latter year, he was called to Rome as director of the Meteorological Office and of the astronomical observatory of the Collegio Romano. It was not until 1887, when the work of the office was extended, that he came into touch with earthquake-investigations in Italy.

After the destruction of Casamicciola in 1883, the Government appointed a royal commission to consider the question of founding a national geodynamic service, the president being Pietro Blaserna (1836–1918), director of the Istituto Fisico in the university of Rome*. The task of the commission was far from simple. In particular their work was hampered by the urgent need for economy, and this was found to be decisive against the creation of a separate earthquake-service. (i) Thus, taking into account the fact that there already existed a network of meteorological stations, at many of which recording instruments had been installed, it was decided that the new service should form a branch of the Central Meteorological Office and should be controlled by the director of that office. (ii) The seismic region, instead of the province, was adopted as the unit-area. Each of the more important regions was to have a central observatory, furnished with the best recording instruments, as well as a large number of stations in which only seismoscopes were to be installed. (iii) A difficult and important problem was the choice of instruments. Of the large number of Italian instruments in use, few could be found to satisfy the required conditions; and it was recognised how far, in this respect, Japan was in advance of Italy. The three-component seismograph and a seismoscope,

* Blaserna was president of the Italian meteorological council from 1878 to 1917. In 1879, he served on the royal commission for the investigation of the Etnean eruption of that year, and, thirty years later, he presided over the royal commission on reconstruction in the areas devastated by the Messina earthquake of 1908 (*Ital. Soc. Sism. Boll.* vol. 22, 1919, pp. 188–201).

designed by Brassart, were, however, recommended for adoption*.

The report of the commission was presented and accepted in 1887, and thereafter the central office was known as the R. Ufficio Centrale di Meteorologia e Geodinamico. As director of the double service, Tacchini was admirably fitted, with his frank and courteous manner, his obvious sincerity, and his persevering energy. He retired from the office in 1899, and died near his birth-place on 24 Mar. 1905†.

113. It is mainly as an astronomer, as a student of solar physics, that Tacchini will be remembered. Until 1887, he had not shown any special interest in seismology—his first note on the subject was written in 1890—but all that he could was done to make the new geodynamic service a success. A special section of the office was created and placed under the charge of his assistant, Giovanni Agamennone. An experimental station was established in the cellars of the office, and observatories for the study of earthquakes were founded at Rocca di Papa, Pavia, Ischia and Catania, as well as others with a wider scope, such as the well-known observatory on Etna.

Not less useful were the journals that Tacchini started, the *Bollettino meteorico* and the *Annali* of the central office, the former with its fortnightly *Supplemento* of earthquake-notices begun in 1889, the latter with its many valuable memoirs on earthquakes. The great advantage of combining the geodynamic service with the meteorological office lay of course in the existing system of reporting stations. Of these, during the first year (1889), not less than 67 were provided with seismoscopes, as a rule, those of Cecchi, Galli and Brassart, 22 possessed seismographs of various Italian patterns, and 12 others had miscellaneous instruments. The value of the *Supplementi* will be evident from the fact that, from 1889 to 1894, the yearly volume contained on an average nearly one hundred quarto pages. During the first four years, the notices related to Italian earthquakes, but, among the records of 1893, are those of the Zante earthquake

* P. Blaserna, Roma, *R. Acc. Lincei Rend.* vol. 4, pt. 1, 1888, pp. 774–782.
† *Ital. Soc. Sism. Boll.* vol. 10, 1904, pp. 169–178.

of 17 Apr. and the Central Asian earthquake of 5 Nov.; among the records of 1894, those of the Hokkaido (N. Japan) earthquake of 22 Mar., the Locris (N.E. Greece) earthquakes of 20 and 27 Apr., the Constantinople earthquake of 10 July, and the Argentine earthquake of 3 Nov. Among the memoirs published in the *Annali* may be mentioned those by T. Taramelli and G. Mercalli on the Riviera earthquake of 23 Feb. 1887 (art. 107), by A. Issel and G. Agamennone on the Zante earthquakes of 1893, and by M. Baratta on the Veronese earthquakes of 7 June 1891 and the earthquakes of 1893 in the Garganic peninsula*.

THE ITALIAN SEISMOLOGICAL SOCIETY

114. The origin of the Italian Seismological Society is to be sought, not in the interest aroused by a disastrous earthquake, so much as in the rapid growth of the earthquake-notices, for which the *Supplementi* proved insufficient, and the slow rate of publication of the *Annali*, the volumes of which were often delayed for two or three years. Tacchini met both difficulties by founding the Società Sismologica Italiana. For such a society, there were ample materials in the country. In the first list of members (1895), we find many well-known names—those, for instance, of the veteran Palmieri, then 87 years old, with only one year more to live, and M. S. De Rossi, who died two years later still (1898). Then comes a group of men who had already achieved fame—Arturo Issel (1842–1922), professor of geology in the university of Genoa, Giuseppe Mercalli (1850–1914), the student of Vesuvius and of the most important earthquake districts of Italy (arts. 103–111), Annibale Riccò (1844–1919), director of the geodynamic observatory of Catania, and Torquato Taramelli (1845–1922), professor of geology in the university of Pavia. Besides these, were a number of younger men destined to take their places a generation later as the leaders of Italian seismology. Salvatore Arcidiacono (1855–1921), assistant in the geodynamic observatory of Catania, in his well-spent life, made many interesting studies of Etna and the Lipari volcanoes, as well as of several Sicilian earthquakes. Mario Baratta, then

* Roma, *Uff. Centr. Meteorolog. Ann.* vol. 8, 1888, 298 pp.; vol. 15, 1893, 200 pp.; vol. 11, 1889, 82 pp.; and vol. 15, 1893, 46 pp.

assistant in the central office of meteorology and geodynamics at Rome, had already begun the regional studies that were to culminate later in his *I Terremoti d'Italia*, the finest monograph that we possess on the earthquakes of any country. Adolfo Cancani (1856–1904), assistant in the geodynamic observatory of Rocca di Papa and afterwards in the central office at Rome, had started the promising career cut short so early. The best work of Giulio Grablovitz, director of the geodynamic observatory of Ischia, of Emilio Oddone (1864–), assistant at the geodynamic observatory of Pavia, of Gaetano Platania, and of Giuseppe Vicentini (1860–), professor of physics in the university of Padua, was yet to come; and only absence from Italy prevented the name of Giovanni Agamennone from appearing in the first list of members. Formerly assistant in the central office at Rome, and in 1895 director of the geodynamic section at Constantinople, he was to succeed De Rossi in the charge of the Rocca di Papa observatory. Of more than a score of volumes of the Society's *Bollettino*, only three were to appear without some work from his pen. It was a band of men from whom much might be expected, and by whom much has certainly been achieved.

The Society differs in more than one respect from those to which we are used in this country. It exists, not so much to promote actual intercourse among its members, as to publish their work on earthquakes and volcanoes without undue delay. By the rules at first drawn up and still in force, the Society was placed under the charge of a director without the assistance of a committee. Apparently almost a dictator, the director is little more than the editor of the Society's journal. From the foundation of the Society, he has been the official head of geodynamic studies in Italy—Pietro Tacchini from 1895 until his death in 1905, and, since then, Luigi Palazzo (1861–), who succeeded him in the direction of the central office at Rome.

115. The first number of the *Bollettino della Società Sismologica Italiana* was published early in 1895, and, by the end of 1924, 24 volumes were completed. As the number of fellows has always been small, never in any year quite reaching one hundred, the size of the volumes is remarkable. The average number of pages in the ordinary text of a volume is 311, and the size of the volumes was more than doubled by the welcome addition of the notices of Italian earthquakes, hitherto issued as a supplement to the *Meteorico Bollettino* of the central office at Rome. The increasing value set on the journal is shown by the growth in the number of foreign or corresponding fellows, from seven in the first year to as many as 44 in 1912, when nearly every country in which earthquakes are studied was represented.

A noticeable feature of the *Bollettino* is the large number of authors. By the end of 1924, papers had been communicated by 54 Italian, and 13 foreign, writers, those of the latter being in English, French and Latin. Among the more important memoirs that have appeared are those by G. Agamennone on the series of Monti Albani earthquakes in Feb. 1906, T. Alippi on brontides, R. de Kövesligethy on the propagation of earthquake-waves, E. Oddone on the recent variations in the sea-level of the central Mediterranean region, G. Platania on the sea-waves of the Calabrian and other earthquakes, and by E. Oddone, A. Cavasino and G. Agamennone on the Marsican earthquake of 13 Jan. 1915*. In addition, though they are outside the range of this volume, may be mentioned the important series of papers on Vesuvian phenomena by G. Mercalli, R. V. Matteucci and others, and those on the state of Etna and the Lipari volcanoes by A. Riccò, S. Arcidiacono, and O. De Fiori.

116. Hardly less important are the notices of Italian earthquakes from 1895 to 1911, comprising more than eight thousand pages. During the war, the issue of the notices was suspended, probably owing to the withdrawal of the government grant. It has since been renewed with the recent publication of the notices for 1911 in a volume of nearly 600 pages, containing records of 664 Italian, and 165 distant, earthquakes†.

* Vol. 21, 1918, pp. 47–101 (Agamennone); vol. 12, 1907, pp. 9–42 (Alippi); vol. 11, 1906, pp. 113–250 (Kövesligethy); vol. 18, 1914, pp. 9–85 (Oddone); vol. 12, 1907, pp. 43–81; vol. 13, 1908, pp. 369–458; vol. 15, 1911, pp. 223–272; vol. 16, 1912, pp. 166–174 (Platania); vol. 19, 1915, pp. 71–217, 219–291; vol. 22, 1918, pp. 9–111 (Marsican earthquake).

† The notices for 1895 were compiled by M. Baratta, 1896 by L. Palazzo, 1897–98 and 1903 by G. Agamennone, 1899–1902 by A. Cancani, 1904–05 by V. Monti, and 1906–11 by G. Martinelli (1877–).

One other point in connexion with the Society should be mentioned here, the first exhibition of seismographs and their records. This was held at the centenary celebration of the Ateneo of Brescia on 6–10 Sep. 1902. The instruments shown were all of Italian origin, including those designed by Agamennone, Cancani, Grablovitz and Vicentini. At the same time and place, the Society held its first meetings, at which these and other instruments were described.

CONCLUSION

117. In the preceding summary of the work done in Italy, I have omitted much that would have deserved mention in a larger volume. No other country has contributed so much to the mass of earthquake-literature. According to Montessus*, we owe to Italian seismologists 2002 out of a total of 8500 memoirs, or very nearly one-quarter, and many of these papers must be inaccessible in British libraries. If we may assume, however, that, in this and the third chapters, the more important work has been referred to, we shall realise how deeply we are indebted to the seismologists of Italy. We owe to them the first scientific investigation of a series of great earthquakes (arts. 26–32), the first observatory erected for the study of volcanic and seismic phenomena (art. 86), the earliest sensitive recording seismoscope (art. 87), the recognition of the reality of microseismic motions (arts. 89–91), the founding of the first seismological journal (art. 99), the device of rational scales of seismic intensity (arts. 100, 101), a series of maps representing the varying distribution of the seismic activity of a country (art. 105), and detailed investigations of some recent great earthquakes (arts. 106–109).

* *Ital. Soc. Sism. Boll.* vol. 20, 1916, pp. 263–272.

CHAPTER VII

THE STUDY OF EARTHQUAKES IN CENTRAL EUROPE

118. In the preceding chapters of this book, the work of early students in Central Europe has been described. Of those in Switzerland, we have Élie Bertrand (1712–*c.* 90) noted for his catalogue of Swiss earthquakes from A.D. 563 to 1754 (arts. 8–11), and Peter Merian (1795–1883) who discovered the greater frequency of earthquakes in the winter months (art. 40). Among German workers have been mentioned K. E. A. von Hoff (1771–1837), the first to compile annual lists of earthquakes and a full catalogue for the whole world (art. 41); P. N. C. Egen (1793–1849), who invented the arbitrary scale of seismic intensity (art. 42); H. Berghaus (1797–1884), the pioneer of earthquake-maps of the world (art. 68); and, lastly, and exceeding them all in power of intellect, A. von Humboldt (1769–1859), in whose *Cosmos* is to be found one of the early attempts to give an account of earthquake-phenomena in general (arts. 45–46).

The present chapter is concerned with the study of earthquakes in Germany, Austria and Switzerland after the year 1845. During the half-century that followed, that study was practically confined to workers in other fields. In Germany, J. F. J. Schmidt was an astronomer; in Austria, L. H. Jeitteles gave most of his attention to ornithology; in Switzerland, F. A. Forel was a many-sided naturalist. With these exceptions, all the prominent investigators of earthquakes were geologists. Moreover, if we may judge from the number of papers published, not one of them, geologist or otherwise, allotted so much as one-tenth of his time to the study of earthquakes. Some of them wrote but one memoir on the subject. Yet—and this is really a matter for regret—the standard attained by these solitary efforts is a remarkably high one.

119. One useful result of the study of earthquakes in Central

Europe during the ten years 1873–82 was the invention of a number of seismological terms.

In 1878, for instance, R. Hoernes divided earthquakes into rock-fall earthquakes (Einsturzbeben), volcanic earthquakes and tectonic earthquakes, a classification that has not yet been superseded. The name centrum or centre for the seismic focus had already been introduced by L. H. Jeitteles in 1859. It was followed, in 1874, by epicentrum or epicentre, for which we are indebted to J. F. J. Schmidt. Stosslinie is the name given by H. Höfer in 1880 to a line that passes through a series of epicentres. Mallet's terms isoseismal line, coseismal line and meizoseismal area were replaced, by K. von Seebach in 1873, by isoseist, homoseist, and pleistoseist zone*. For the strongest member of a series of earthquakes, the natural term principal shock was suggested by J. F. J. Schmidt in 1874. The slighter shocks of the series were called accessory shocks by Forel in 1881, those before the principal shock being preparatory shocks and those after it consecutive shocks. For these, R. Canaval in 1882 devised the useful terms Vorbeben and Nachbeben, terms that were to reappear later and independently as fore-shocks and after-shocks.

THE STUDY OF EARTHQUAKES IN GERMANY

120. Few men, in the course of a long life, can have devoted themselves so entirely to scientific pursuits as Johann Jakob Nöggerath (1788–1877), professor of mineralogy and mining in the university of Bonn. His first paper was published in 1811, his last in 1873. Most of his memoirs deal with geological subjects, but, from 1828 to 1870, he wrote as many as thirteen papers on earthquakes. Two of them are of especial interest, one dealing with the Rhenish earthquake of 29 July 1846, the other with the Rhenish earthquakes of the years 1868–70.

The most important of Nöggerath's seismological memoirs is that on the Rhenish earthquake of 29 July 1846†. It is notable

* Seebach's terms are more correctly formed than Mallet's, and both homoseist and pleistoseist have passed into current use. His terms "Erdbebenursprungsort" and "Oberflächenmittelpunkt," instead of centre and epicentre, have not been accepted.

† *Das Erdbeben vom 29 Juli 1846 im Rheingebiet und den benachbarten Ländern*, Bonn, 1847, 60 pp. and 1 map.

especially on two accounts, the construction of the earthquake-map and the estimate of the velocity of the earth-wave. (i) So far as I know, the earthquake-map contains the first attempt to draw isoseismal lines, very roughly no doubt, and leading to the first determination of the position of the epicentre by means of

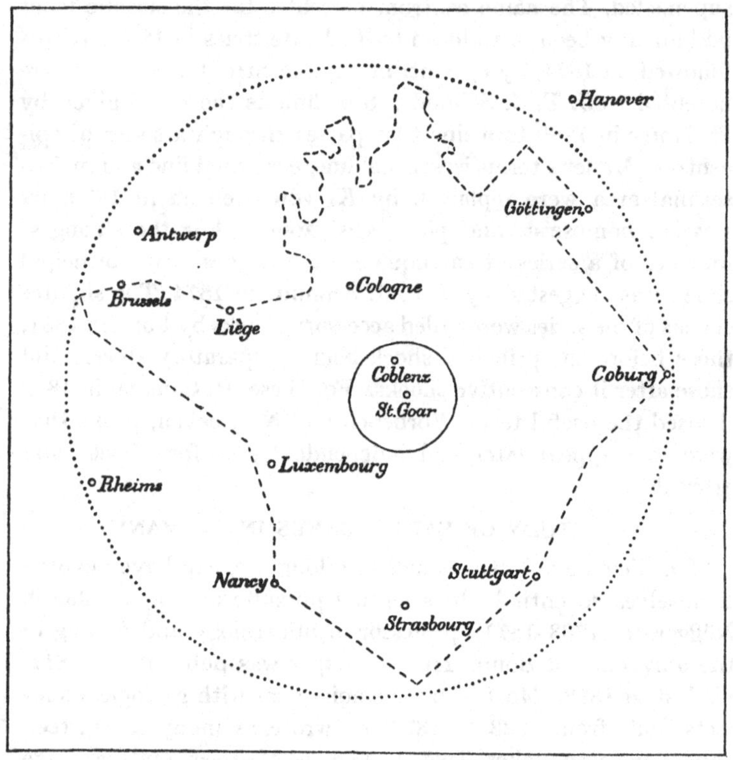

FIG. 6. Rhenish earthquake of 29 July 1846; map of isoseismal lines.

such lines. The essential features of the map are shown on a smaller scale than the original in Fig. 6. The polygonal broken-line joins the outermost places at which the shock was actually felt, and the dotted circle, which just surrounds these places, indicates the probable form of the disturbed area. The small continuous circle near the centre of the area bounds the district in which the shock was felt with the greatest intensity. The radius of this circle is 28 miles, the area within it about 2500 square miles, and the centre is close to the small town of St Goar on the Rhine. (ii) Not less interesting is the determination of the velocity of the earth-wave. This section is the work of J. F. J. Schmidt (1825–84; art. 126), at that time an assistant in the observatory of Bonn. Taking St Goar as the centre, Schmidt found the mean velocity to be 3·684 Prussian miles a minute or 1517 ft. per sec. Rough estimates of the velocity had been made for previous earthquakes—for instance, that of Michell of more than 20 miles a minute for the Lisbon earthquake (art. 16)—but this estimate of Schmidt's is probably the first with any pretence to accuracy, certainly the first in which it was determined by the method of least squares.

Nöggerath's last memoir on earthquakes* is a remarkable one, seeing that it was written at the age of 82. It contains somewhat detailed accounts, though without maps, of six Rhenish earthquakes during the years 1868 and 1869. Lists are given of the remarkable series of slight earthquakes felt in the Grand Duchy of Hesse in 1869 and 1870, especially at Gross-Gerau, where as many as 223 shocks were felt in three weeks. The opportunity was also taken to give a catalogue of earthquakes in Rhineland, 270 in number, beginning with the year 801 and ending with 1858.

121. Georg Heinrich Otto Volger (*c.* 1822–97), the historian of Swiss earthquakes, was in middle life a notable figure in scientific and political circles in Frankfort. Except for a brief paper on the earthquake of 29 July 1846, his first seismological article was a preliminary account of the Visp valley earthquake of 25 July 1855†. The map that illustrates this paper‡ is probably the first on which isoseismal lines depending on a definite scale of intensity were depicted. The five curves drawn correspond to the following intensities: (1) whole houses and churches fell; (2) the strongest parts of buildings damaged; (3) slight injuries

* Bonn, *Nat. Hist. Ver. Verh.* vol. 27, 1870, pp. 1–132.

† Petermann, *Mitth.* 1856, pp. 85–102. It may be of interest to recall that this earthquake was the subject of a brief paper by Osmond Fisher (1819–1914; *Phil. Mag.* vol. 11, 1856, pp. 240–242) in which, for the first time, I believe, an earthquake was attributed to the growth of a fault.

‡ Drawn by A. Petermann (1822–78) from materials supplied by Volger.

to buildings; (4) little damage to buildings; (5) the shock certainly felt. As will be seen from Fig. 7, the curves are extremely irregular in form.

This paper was the forerunner of Volger's most important work, that on the earthquakes of Switzerland*. The earthquake-

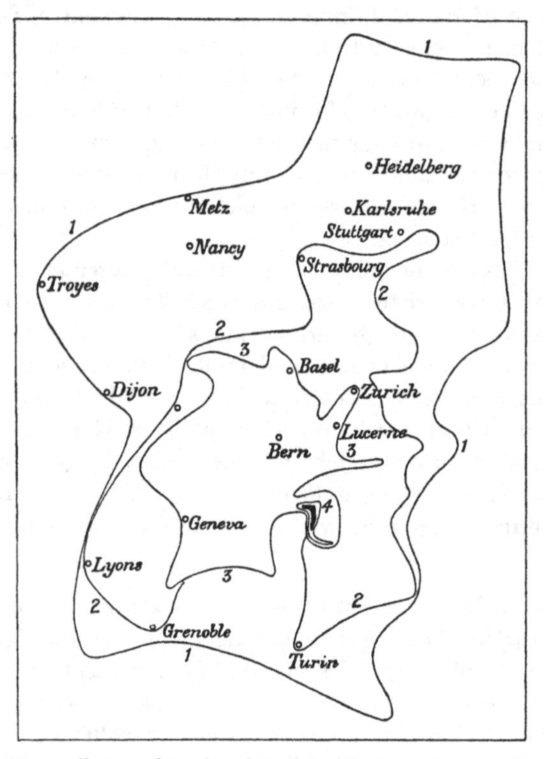

Fig. 7. Visp valley earthquake of 25 July 1855; map of isoseismal lines.

chronicle, which occupies the first part of the work, contains records of 1347 earthquakes. Of 1230 of these, the month of occurrence is known, and their distribution throughout the year is as follows:

Jan.	Feb.	Mar.	Apr.	May	June	July	Aug.	Sep.	Oct.	Nov.	Dec.
150	143	138	119	58	54	40	47	117	111	85	168

* *Untersuchungen über das Phänomen der Erdbeben in der Schweiz*, the first part (367 pp., 1857) being a chronicle of earthquakes in Switzerland,

or 456 earthquakes during the summer half of the year and 774 during the winter half. Similarly, of 435 earthquakes of which the hour of occurrence is known, the figures show the usual observed preponderance during the night hours, being:

	0–2	2–4	4–6	6–8	8–10	10–12
a.m.	49	55	43	31	31	36
p.m.	18	27	31	24	42	48

or 152 from 10 p.m. to 4 a.m., 105 from 4 a.m. to 10 a.m., 81 from 10 a.m. to 4 p.m., and 97 from 4 p.m. to 10 p.m.

The whole of the third part, a volume of more than five hundred pages, is devoted to the Visp valley earthquake of 25 July 1855, its predecessors from Dec. 1854 and its successors to Nov. 1855. During this year, the distribution of shocks throughout the day resembles that obtained from the chronicle of previous earthquakes, being 156 from 10 p.m. to 4 a.m., 91 from 4 a.m. to 10 a.m., 99 from 10 a.m. to 4 p.m., and 107 from 4 p.m. to 10 p.m. The velocity of the earth-wave was estimated at 2984 ft. per sec., which agrees closely with that obtained for the Hereford earthquake of 1896, namely, 2955 ft. per sec. In the map that accompanies this part, the isoseismal lines (seven in number) differ rather widely from those given in the early report, and the boundaries of the areas disturbed by three important after-shocks (on 26 and 28 July) are also shown.

122. Three years after the publication of Nöggerath's last paper on earthquakes, there began the most active period of German seismology during the latter half of the nineteenth century. The leading memoir was Seebach's report on the mid-German earthquake of 1872, followed by the memoirs of Lasaulx on the Herzogenrath earthquakes of 1873 and 1877, of Schmidt on the earthquakes of south-eastern Europe, and of Geinitz on the sea-waves of the Iquique earthquake of 1877. Almost concurrently, there was a similar period of work on earthquakes in Austria, during which reports were written by Bittner and Höfer on the Belluno earthquake of 1873, by Wähner and Hantken on the Agram earthquake of 1880, by Canaval on the

the second (283 pp., 1857) an outline of the geology of canton Valais, and the third (524 pp., 1858) a report on the earthquakes of the same canton during the year 1855.

Gmünd earthquake of 1881, and, after a long interval, by F. E. Suess on the Laibach earthquake of 1895. Within the same period, we have also the remarkable memoirs of E. Suess and Hoernes. In Switzerland, the renewal of activity dates mainly from 1878 with the foundation of the Swiss Seismological Commission.

123. Of great performance and still greater promise was Karl Albert Ludwig von Seebach (1839–80), professor of geology in the university of Göttingen. Though young in years—he died at the age of forty—he had travelled in Costa Rica (1864–65) and the Aegean Sea (1866), and, in addition to many works on geology, had written memoirs on the volcano of Santorin and its eruption in 1866, on the volcanoes of Central America and on the mid-German earthquake of 6 Mar. 1872*.

Seebach's single contribution to seismology† is especially notable for the use that he made of time-observations. He drew two maps of the earthquake, one showing the pleistoseist zone and the isoseists, and the other the homoseists (art. 119). In the former, the pleistoseist zone is represented as a narrow band 15 miles long and lying about 30 miles south of Leipzig. Of the three curves, which he calls isoseists, one bounds the area within which damage occurred to buildings, the second that in which the accompanying sounds were observed, and the third marks the limits of the disturbed area. Though Mallet had noticed in 1862 how the sound-observations during the Neapolitan earthquake of 1857 were crowded within a small area surrounding the epicentre, this map of Seebach's is, I believe, the first in which the boundary of the sound-area is definitely traced. It will be noticed that he regarded it as an isoseismal line.

Seebach's faith in his time-observations led to one curious result. The times, to the nearest minute, were 3·58 p.m. at Eger and Halle, and 3·59 at Göttingen and Leipzig. Drawing straight lines joining each pair of places, their perpendicular bisectors intersect in a point which he regarded as the epicentre. He is careful to give its position rather minutely, as 50° 38′·6 N. lat.,

* *Neues Jbuch. Min.* 1880 (pt. 1), 8 pp.; *Allg. Deutsch. Biog.* vol. 33, 1891, pp. 557–559.

† *Das mitteldeutsche Erdbeben vom 6 März* 1872, 1873, 192 pp., 5 plates.

11° 1′·25 E. long. Strangely enough, this point lies not only outside the pleistoseist zone, but even outside the area of damage. It is, indeed, 50 miles from the nearer end of the one and 25 miles from the boundary of the other. How excentrically it is placed is evident from the fact that the second isoseist passes 12 miles to the southwest, and more than 150 miles to the north-east, of it. The determination, however, receives some support from the homoseists drawn for every minute from 3·58 to 4·5, these curves being circles with the ascertained point as centre.

124. The most original part of Seebach's memoir is his method of determining the depth of the focus from observations of the absolute time. In Fig. 8, distances from the epicentre in miles are measured from *O* along *Ox*, and *OM* represents the distance of a given place of observation. The number of minutes elapsed since a given time (in this case, 3.55 p.m.) are measured parallel to *Oy*, and *MP* represents the number of minutes between 3.55 and the observed time at the place. With two permissible simplifications, Seebach shows (i) that *P* and all similar points lie on the hyperbola *EP*, (ii) that the cotangent of the angle between the asymptote *FQ* and the line *Ox* gives the velocity of the earth-wave (supposed constant) in the outer crust, (iii) that the lengths *OF* and *OE* determine the times at the focus and epicentre, respectively, and (iv) that, since the velocity is known, the interval represented by *FE* permits the depth of the focus

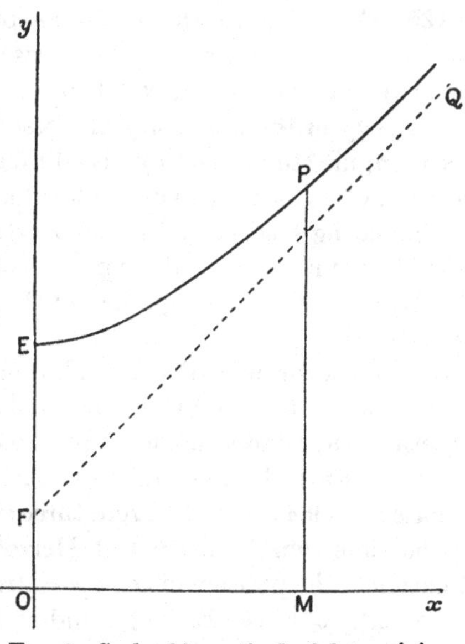

FIG. 8. Seebach's method of determining the depth of the seismic focus.

to be determined. The method, he says, is so clear, convenient and sure, that he hopes that it will be used whenever an earthquake visits a civilised country*.

Applying it to the mid-German earthquake, Seebach found that the mean velocity of the earth-wave was 2430 ft. per sec., that the earthquake occurred at the focus at 3h. 56m. 9s. p.m., Berlin time, and that the depth of the focus was about 12 miles.

For two other earthquakes—the Rhenish earthquake of 29 July 1846 (art. 120) and the Sillein earthquake of 15 Jan. 1858 (art. 129)—Seebach had materials for the use of his method. For these, he found the velocities of the earth-wave to be about 2050 and 750 ft. per sec., and the depths of the foci to be about 24 and 16 miles.

125. Following in the footsteps of Seebach as closely as Oldham and Johnston-Lavis in those of Mallet (arts. 81–83), came Arnold von Lasaulx (1839–86), afterwards professor of mineralogy in the university of Breslau. Widely known by his petrographic studies and by his edition of the posthumous work on Etna of A. Sartorius von Waltershausen, remembered also as a stimulating teacher and a many-sided investigator†, Lasaulx is worthy of mention in these pages as the author of memoirs on the Herzogenrath earthquakes of 22 Oct. 1873 and 24 June 1877‡.

(i) For the former earthquake, Lasaulx' map closely resembles Seebach's of the mid-German earthquake. He gives three isoseists—the boundaries of the pleistoseist zone, the sound-area, and the disturbed area—and also a series of homoseists for every minute from 9.42 to 9.47. From three simultaneous observations of the time (9h. 41m. 33s.) at Herzogenrath, Kumpchen and Richterich, the position of the epicentre is determined as 50° 52′ 51″ N. lat., 6° 3′ 15″ E. long., and it is worthy of notice that

* I have described Seebach's method somewhat fully on account of its theoretical interest. It is doubtful, however, whether the time-records are sufficiently accurate to bear the strain that is placed on them.

† *Neues Jbuch. Min.* 1886 (pt. 1), 6 pp.; *Allg. Deutsch. Biog.* vol. 51, 1906, pp. 595–596.

‡ (i) *Das Erdbeben von Herzogenrath am 22. October* 1873, 1874, 157 pp., 4 plates; (ii) *Das Erdbeben von Herzogenrath am 24. Juni* 1877, 1878, 77 pp., 1 plate. Herzogenrath lies about 40 miles west of Cologne.

this point is close to the centre of the pleistoseist zone. The time-distance curve, constructed as in Seebach's method, then gives the velocity of the earth-wave as 920 ft. per sec., the time of occurrence at the focus as 9h. 41m. 30s., and the depth of the focus as about 7 miles.

(ii) The earthquake of 1877 is treated in somewhat less detail and no map of the earthquake is given. Following Seebach's method again, Lasaulx found the epicentre to be in 50° 52′ 51″ N. lat., 6° 1′ 51″ E. long. The time-distance curve gives 1560 ft. per sec. for the velocity of the earth-wave, 8h. 52m. 12s. for the time of occurrence at the focus, and about 17 miles for the depth of the focus.

126. It is chiefly as an astronomer that Johann Friedrich Julius Schmidt (1825–84) will be regarded, by his studies on variable stars, sun-spots, comets and the zodiacal light, but above all by his elaborate map of the moon published in 1878. At the age of twenty, he began to collect materials for a world-catalogue of earthquakes, and, in the following year, he contributed the estimate on the velocity of the earth-wave to Nöggerath's memoir of the Rhenish earthquake of 1846 (art. 120). His appointment in 1858 as director of the observatory of Athens brought him within the range of the earthquakes of south-eastern Europe. Among his works are included volumes on the eruption of Vesuvius in 1855 and *Vulkanstudien*. He wrote two memoirs on earthquakes, one on the slightly destructive Sillein (Hungary) earthquake of 15 Jan. 1858, the other his well-known volume *Studien über Erdbeben*. Hampered by the lack of fine instruments, Schmidt's life is nevertheless a record of what may be done by boundless enthusiasm and unresting labour*.

By seismologists, Schmidt is valued for his admirable *Studien über Erdbeben*†, probably the only original work on earthquakes that has ever passed into a second edition. As already noticed (art. 119), it was in this memoir that he introduced the most useful term *epicentre*. Of the three parts, into which the volume is divided, the first (pp. 1–34) deals with the frequency of earth-

* *Astr. Soc. Month. Not.* vol. 45, 1885, pp. 211–218; *Observatory*, London, vol. 7, 1884, pp. 118–119; Meyer's *Konv. Lex.* vol. 17, 1909, p. 897.

† First edition, 1874; second edition, 1879, 360 pp., 6 plates.

quakes in connexion with various astronomical and meteorological phenomena; the second (pp. 35–136) contains monographs of 23 earthquakes in the south-east of Europe from 1837 to 1873; while the third (pp. 137–360) is a catalogue of earthquakes in the same region from the earliest times to 1878. Taking them in the inverse order, the catalogue up to 1858 is founded chiefly on the lists of Perrey and Mallet; from 1859 onwards, it is the result of Schmidt's own labours. Altogether, for this one region, the total number of entries amounts to more than four thousand, nearly nine-tenths of them dating from after 1858. Of the earthquakes described in the second part, the most detailed accounts are given of the Aegian earthquake of 26 Dec. 1861, the Cephalonian earthquake of 4 Feb. 1867, and the Phocis earthquake of 1 Aug. 1870. These disturbed areas of 25,000, 125,000 and 50,000 sq. miles, respectively, amounts that were far exceeded in the Mediterranean earthquake of 24 June 1870, of which the disturbed area was about $1\frac{3}{4}$ million sq. miles.

Schmidt's interests naturally lay in the astronomical or meteorological relations of the earthquakes. He examined first their lunar periodicities. His curve representing the frequency of earthquakes in connexion with the relative positions of the sun and moon can hardly be said to agree with Perrey's first law, for it gives maxima about the time of new moon and two days after first quarter. Perrey's second law is, however, confirmed, earthquakes being more frequent when the moon is in perigee than when it is in apogee (art. 56–58). As regards their seasonal distribution, there were, from 1200 to 1873, 564 earthquakes in the three winter months (Dec.–Feb.), 574 in spring, 418 in summer, and 622 in autumn. The apparent greater nocturnal frequency receives strong support, as 1281 earthquakes are recorded from 6 p.m. to 6 a.m. and 891 from 6 a.m. to 6 p.m. Lastly, earthquakes are found to occur more frequently with a low, than with a high, barometer.

127. Within the scope of this section, as it belongs to the year 1878, must be reckoned the work of Franz Eugen Geinitz (1854–), professor of mineralogy and geology in the university of Rostock. His single, but important, contribution to seismology is a study of the sea-waves of the Iquique earthquake of 9 May 1877*. These were observed at more than forty stations along the west coast of South America for a distance of 2200 miles. The most interesting records, however, are those which come from great distances, from the Hawaiian Islands (6500 miles from the epicentre), Samoa (6600 miles), New Zealand (5600 miles), New South Wales (6800 miles) and Japan (8900 miles). The total number of such records collected by Geinitz amounts to 12. In each case, he calculated the mean velocity of the sea-waves, and thence the corresponding mean depths of the Pacific Ocean along the paths taken by the waves. For this he used the formula $h=v^2/g$†, which applies to the case when the height of the waves is small compared with the depth of the ocean, h being the mean depth in feet and v the mean velocity of the waves in feet per second. The mean depth was thus found to be 1930 fathoms to Samoa, 2322 fathoms to Honolulu, 1423 fathoms to Wellington (N.Z.), 2064 fathoms to Sydney, and 2181 fathoms to Kamaishi (Japan). While some of these depths agree roughly with those calculated from Petermann's chart of the Ocean, Geinitz notices that others are distinctly inferior. A few years later, Milne discussed the same and other observations and remarked on the same inequality. He suggested that it might be due to over-estimates in the sounded depths (art. 195). It has, however, been shown that an inequality of the kind and order observed is a necessary result of variations in depth along the path traversed by the waves‡.

THE STUDY OF EARTHQUAKES IN AUSTRIA

128. First among Austrian geologists to devote himself temporarily to seismology was Ferdinand von Hochstetter (1829–84), "an admirable geologist and an indefatigable worker." Having joined the geological survey of Austria in 1852, he took part in the *Novara* expedition from 1857 to 1859. In the latter year, on arriving in New Zealand, he made a rapid survey of the

* *Ac. Nat. Curios. Nova Acta*, vol. 40, 1878, pp. 383–444.

† Following Hochstetter (art. 128), Geinitz gives two expressions for the above formula, $h=(v/5{\cdot}671)^2$ and $h=v^2/g$, results which are in reality the same as 5·671 is the square root of g.

‡ *Phil. Mag.* vol. 50, 1900, pp. 583–584.

colony, the outcome of which were volumes on its geology and paleontology. He returned to Vienna in 1860, becoming professor of geology in the Royal and Imperial Polytechnic Institute and afterwards superintendent of the State Natural History Museum*. It was while he held the former post that he wrote the three parts of his valuable memoir on the sea-waves of the Peruvian earthquake of 13 Aug. 1868†.

The centre of this great earthquake was near Arica on the coast of Peru, and Hochstetter collected observations of the waves from many stations on the coast, from the Chincha Islands on the north to near Valdivia on the south, places that are 2000 miles apart. The waves were also recorded at many distant harbours, as far as the Hawaiian Islands (6218 miles), the Chatham Islands (6356 miles), Samoa (6633 miles), New Zealand (7047 miles), and New South Wales (8500 miles). From the velocities along each of these paths, Hochstetter calculated the corresponding mean depths of the ocean by means of the formula $h=v^2/g$ (art. 127). His results are 2266 and 2882 fathoms to the Hawaiian Islands, 1912 fathoms to the Chatham Islands, 1891 fathoms to Samoa, 1473 fathoms to New Zealand, and 1501 fathoms to Newcastle in Australia. It will be seen that these figures are of the same order of magnitude as those obtained nine years later by Geinitz from the Iquique earthquake of 1877 (art. 127).

129. Of the strong Austrian earthquakes in the latter half of the nineteenth century, one of the most important was the Sillein earthquake of 15 Jan. 1858, studied by J. F. J. Schmidt (art. 126) and by Ludwig Heinrich Christian Jeitteles (1830–83). Jeitteles' map of the earthquake contains three isoseismal lines which bound the areas in which the shock was strongest, in which it was distinctly and almost universally felt, and in which it was weak and felt by a small proportion of the inhabitants. It shows very clearly the influence of the form and structure of the ground on the intensity of the shock. He points out that it

* *Geol. Soc. Quart. Jl.* vol. 41, 1885, p. 45 (*Proc.*); *Geol. Mag.* vol. 3, 1884, pp. 526–528.

† Wien, *Ak. Sber.* vol. 58, 1868, pp. 837–860; vol. 59, 1869, pp. 109–132; vol. 60, 1870, pp. 818–823.

was only in the valleys and on low ground that the shock was strongly felt; on higher land, it was perceived but slightly or not at all. Again, the shock was more intense on recent rocks, and slight on the harder crystalline masses*.

130. A brief note on the same earthquake was also written by Dionys Stur (1827–93), then a member of the Austrian geological survey, of which he afterwards became director. A more important contribution was his report on a remarkable series of earthquakes at Klana, near Fiume, mostly during the first half of the year 1870†. In this memoir, no new methods of study were introduced, but it contains some interesting observations on the brontides observed in the district.

131. The Belluno earthquake of 29 June 1873 was one of some consequence, for, by it, 41 persons were killed and more than 2500 houses were seriously damaged. One of the first to study it was the Austrian geologist Alexander Bittner (*c.* 1850–1902). His valuable report‡ is of much statistical interest. In addition to the detailed survey of the effects produced by the shock, he gives two useful tables, one of about 140 earthquakes in the district from A.D. 375 to 1873, the other of more than 100 accessory shocks of the earthquake of 1873 from 22 June to 25 Dec.

132. Three years later, another report on the Belluno earthquake was published by Hans Höfer (1843–), professor of geology at Leoben. His study§ is notable on two accounts. (i) It contains the suggestion, I believe the first suggestion, that an earthquake may originate in two distinct foci. The principal epicentre, close to the village of Quantin, he takes to be the centre of a circle passing through the four places that suffered most from the earthquake. A secondary epicentre is placed near Corni, about 5 miles to the east of the other. (ii) Like Seebach, Höfer puts great, I think too great, trust in observations of the time, but the result at which he arrives is certainly remarkable. With the

* Wien, *Ak. Sber.* vol. 35, 1895, pp. 511–592.
† Wien, *Geol. Jbuch.* vol. 21, 1871, pp. 231–264.
‡ Wien, *Ak. Sber.* vol. 69, 1874, pp. 541–637.
§ Wien, *Ak. Sber.* vol. 74, pt. 1, 1877, pp. 819–856.

aid of 13 time-records in the neighbourhood of 5 h. 0 m., he draws the homoseist corresponding to this epoch. The main part of it is an oval curve, 200 miles long, with its axis running north-west and south-east. About the middle of the east side, a secondary portion, about 75 miles in length, branches off abruptly towards the east. The form of the homoseist he regards as similar to that of the focus, and he states that the axis of both portions of the curve are occupied by prominent faults, which intersect in the neighbourhood of the principal epicentre*.

To Höfer, we are also indebted for a memoir on the earthquakes of Carinthia†, of which he describes about 200 from the close of the eighth century. The interesting feature of the memoir is the series of lines—Stosslinien—that he draws passing through or near the epicentres of these earthquakes. Of these lines, the more prominent have a general easterly or south-easterly direction.

133. Among Austrian geologists, past and present, none holds so high a place as Eduard Suess (1831–1914), professor of geology at Vienna from 1857 to 1901. His principal contributions to seismology are two memoirs, both written in 1873, on the earthquakes of Lower Austria and of Southern Italy, and the chapter on some seismic areas in the first volume of his great work *Das Antlitz der Erde* (1885–1909). Thus, his total output amounts to little more than a hundred pages, and seismologists can only regret that, after 1885, his attention was diverted from earthquakes by other and more pressing work.

(i) Many of the earthquakes in Lower Austria‡ have attained destructive strength, and they are of special interest from the fact that the geology of the country is so well known. Suess refers in particular to the strong earthquakes of 15–16 Sep. 1590, 27 Feb. 1768 and 3 Jan. 1873, but he also gives a list of more than 120 earthquakes in the district from 1021 to 1873. He shows that, with a few exceptions of little consequence, the

* It should be mentioned that Höfer's results are contested by R. Hoernes (Wien, *Geol. Jbuch.* vol. 28, 1878, pp. 409–420), who denies the existence of the faults referred to.

† Wien, *Ak. Denkschr.* vol. 42, pt. 2, 1880, pp. 1–90.

‡ Wien, *Ak. Denkschr.* vol. 33, pt. 1, 1874, pp. 61–98.

earthquakes are connected with transversal faults or flaws, and that shocks of varying intensity have repeatedly occurred along the same lines of fracture.

(ii) Of still higher interest are the earthquakes of Southern Italy considered in Suess' second memoir*. He assigns the remarkable series of 1783 to movements along a peripheral fracture traversing Calabria from the west side of Aspromonte and passing close by Reggio and Etna. The fracture is roughly a circular arc with a radius of about 60 miles, lying to the east and south of the Lipari Isles. It is crossed by six radial seismic lines converging nearly towards the same islands, lines that are beset with volcanoes near their junction. The crust within the peripheral line he supposes to have sunk down in the form of a dish, and this subsidence would result in the production of radial fractures. Any disturbance in the equilibrium of the blocks so formed would give rise to increased volcanic activity and to earthquakes in the mainland and in Sicily.

(iii) The four seismic areas that Suess considers in the *Antlitz*† are the north-eastern Alps, the south of Italy, Central America, and the western coast of South America. For the first two areas, he summarises his own early work described above. The last section is devoted to a criticism of the reported elevations of the Chilian coast during the earthquakes of 1822, 1835 and 1837. For the two earlier earthquakes, the evidence has been considered in a former chapter (arts. 33–39). Suess, as is well known, preferred the negative evidence, and, brushing aside the other, concluded that the observations of elevation were erroneous. Whether his scepticism in these particular cases was justified or not—and, in my opinion, it was not justified—is now of little moment, for the question of elevation has been fully settled by later events, and especially by the Alaskan earthquakes of 1899 and the Japanese earthquake of 1923.

134. Rudolf Hoernes (1850–1912) was a pupil of Eduard Suess and afterwards professor of geology at Graz, the capital of Styria. To seismologists, he is well known by his memoir on

* Wien, *Ak. Denkschr.* vol. 34, pt. 1, 1875, pp. 1–32.
† *The Face of the Earth*, vol. 1, 1904, pp. 73–105.

Erdbeben-Studien. He also wrote *Erdbebenkunde* (a textbook of seismology), a criticism of the earthquake-theories of Rudolf Falb (1838–1903), and reports on several Styrian earthquakes and on the Greek earthquakes of 1902 and 1904.

In the *Erdbeben-Studien**, Hoernes introduced his useful classification of earthquakes into rock-fall earthquakes, volcanic earthquakes and tectonic (afterwards dislocation) earthquakes. Realising that rock-fall earthquakes are infrequent and volcanic earthquakes local, he holds that the most numerous, as well as the greatest, of all earthquakes are the products of mountain-formation, and are due to displacements along peripheral and radial fractures of great mountain-chains. Hoernes' studies are mainly based on the Belluno earthquake of 1873, the Klana earthquakes of 1870, and the Villacher earthquake of 1838. From these and other earthquakes felt in Austria and Germany, he draws most of the illustrations in his textbook *Erdbebenkunde*†. This natural preference and the frequent quotation from writers in Central Europe—more than two-thirds of the references are to memoirs written in German—increase the usefulness of the book to readers in other lands.

135. The destructive Agram earthquake of 9 Nov. 1880 was the subject of many memoirs, two of which must be referred to here. The Hungarian geologist Max Hantken von Prudnik (1821–93), was chiefly concerned with estimating the amount of the damage and the value of the property destroyed‡, though he also gives lists of the Agram earthquakes from 1502 to 1880 and of the after-shocks of the last earthquake up to 4 Mar. 1881. Franz Wähner (1856–), professor of geology at Prague, also made an exhaustive study of the effects of the earthquake§. Like his predecessor, he gives a list, but rather fuller, of Agram earthquakes beginning in the same year (1502), and another of the accessory shocks from 9 Nov. 1880 to 21 Jan. 1882. His map of the earthquake, which is the more detailed of the two, depicts four areas, namely, those of important destruction of buildings, strong, slight, and no, damage.

* Wien, *Geol. Jbuch*. vol. 28, 1878, pp. 387–448. † 1893, 452 pp.
‡ *Ung. Geol. Anstalt Jbuch*. vol. 6, 1882, pp. 47–132.
§ Wien, *Ak. Sber*. vol. 88, pt. 1, 1884, pp. 15–344.

136. A few years after the retirement of Eduard Suess in 1901, he was followed in the chair of geology at Vienna by his son Franz Eduard Suess (1867–), whose principal contribution to seismology is his elaborate account of the Laibach earthquake of 14 Apr. 1895*. This memoir is illustrated by two maps, one of the pleistoseist region, the other of the disturbed area. On the former, places are indicated by six different symbols according to the amount of damage caused by the earthquake, the corresponding intensities forming a scale for moderately destructive earthquakes. On the second map are shown seven isoseismal lines, four of which represent varying amounts of damage, while the rest correspond to the effects on the population—general perception of the shock and the predominance of positive or negative reports. The memoir is notable for the unusual attention paid to the sound-phenomena and the observation of the earthquake in mines, as well as for the attempt to construct the hodograph† or time-distance curve. The earthquake was one of the early group registered instrumentally at many points outside the disturbed area, as far as Ischia (371 miles from the epicentre), Grenoble (428 miles), Potsdam (450 miles) and Wilhelmshaven (596 miles). It resembles others of the group in the somewhat discordant estimates of the velocity along different paths, due chiefly to the variable sensitiveness of the instruments employed‡.

137. For about twenty years, beginning with 1865, Carl Wilhelm C. Fuchs (*c.* 1837–86) published annual lists of volcanic eruptions. Otherwise, his papers are mostly on geological subjects, though they include four on earthquakes. Of these, by far the most useful was his long memoir containing lists of earthquakes for the twenty years 1865–84§. These lists are given for 44 districts, and include altogether nearly ten thousand entries. They suffer, in common with all

* Wien, *Geol. Jbuch.* vol. 46, 1897, pp. 411–614.

† For the term *hodograph*, we are indebted to A. Schmidt (*Württemb. Jhefte*, 1888, pp. 248–270; 1900, pp. 200–232).

‡ See, for instance, art. 163 and *Brit. Ass. Rep.* 1894, pp. 146–154.

§ *Statistik der Erdbeben von* 1865–85. Wien, *Ak. Sber.* vol. 92, pt. 1, 1886, pp. 215–625.

world-catalogues, from the difficulty of appreciating the value of records from foreign countries*.

SEISMOLOGICAL COMMITTEES IN SWITZERLAND AND AUSTRIA

138. Swiss Seismological Commission. Until this Commission was founded in 1878, no earthquake committee had survived for more than a few years. The earliest known to me is the group of workers who joined in the study of the Chichester earthquakes of 1833†. It was followed by two committees of the British Association, both appointed to study the earthquakes of the United Kingdom, but practically confined to those of Comrie in Perthshire. The first, with David Milne (1805–90; art. 44) as secretary, presented four reports in 1841–44; for the second, the secretary James Bryce (1806–77) wrote seven brief reports in 1870–76. In Switzerland, the materials for investigation are far more ample than in this country, and the Swiss Seismological Commission and its successor, the Swiss Earthquake Service, have carried on their work until the present day. For the foundation of the earlier body, we are indebted to the well-known geologist Albert Heim (1849–), who was joined by F. A. Forel (1841–1912), A. Forster and others; for the continuance of its useful work mainly to A. de Quervain of Zurich. The Swiss Seismological Commission was not only the first of its kind to endure; it has served as the model of similar committees in other countries.

139. Albert Heim (1849–), professor of geology in the university of Zurich, was already distinguished in 1878 by his researches on the structure of the Alps. It was in this year, when he was not yet thirty years of age, that his great work *Mechanismus der Gebirgsbildung* was published. In the preceding year, he had made a careful study of the earthquake of 2 May 1877, felt over the greater part of Switzerland, and it was this study that led to his proposal to the Helvetic Society of Natural Sciences that a commission should be appointed for the investi-

* For instance, of 75 disturbances entered for Great Britain, only 34 seem to have been true earthquakes.

† *Roy. Soc. Proc.* vol. 3, 1887, p. 338.

gation of all Swiss earthquakes. In addition to the annual reports prepared by Heim, mention should be made of the useful handbook on earthquakes and their study, which he wrote on behalf of the Commission*.

140. François Alphonse Forel (1841–1912) was one of the last of the race of naturalists who took all science for their province. He was born at Morges on the north shore of the Lake of Geneva, and in 1870 was appointed professor of anatomy and physiology in the neighbouring university of Lausanne. This office he held until 1895, when he retired in order to obtain more time to pursue the studies with which his name is identified.

Forel's published papers are more than 300 in number. His efforts were, however, concentrated on three principal subjects: on the glaciers of Switzerland, their periodic variation, and the structure of the glacier-grain; on the seiches of the Swiss lakes —he has been called the "Faraday of seiches"—and on the lake of Geneva, along the shores of which most of his life was spent. His great work *Le Leman* appeared in three volumes in 1892, 1896 and 1905. In the study of this lake, it has been said, he was in turn geographer, geologist, meteorologist, engineer, chemist, zoologist, botanist, archaeologist, historian and economist†.

141. On its foundation in 1878, the Swiss Seismological Commission consisted of seven members, A. Forster, of the observatory of Bern, being president, and A. Heim secretary. To each member was allotted the task of collecting observations from two or more cantons. Zurich, Uri, etc., were assigned to Heim, Vaud, Valais and Neuchâtel to Forel, Bern and Fribourg to Forster, and so on.

During the early years of its existence, the annual reports of the Commission were published in the *Jahrbücher* of the observatory of Bern. Forel also made the work of the Commission widely known by his four valuable memoirs on the earthquakes of 1880–86‡. In the first of these (pp. 465–466), he gave his scale

* *Die Erdbeben und deren Beobachtung*, Basel, 1880, 31 pp.; translated into French in *Arch. Sci. Phys. Nat.* vol. 3, 1880, pp. 261–289.

† *Arch. Sci. Phys. Nat.* vol. 34, 1912, pp. 185–188; *Geol. Soc. Quart. Jl.* vol. 69, 1913, pp. lvii–lx; *Nature*, vol. 89, pp. 638–639.

‡ *Arch. Sci. Phys. Nat.* vol. 6, 1881, pp. 461–494; vol. 11, 1884, pp. 147–182; vol. 13, 1885, pp. 377–396; vol. 19, 1888, pp. 39–66.

of intensity, afterwards combined with the De Rossi scale to form the Rossi-Forel scale (art. 101). He also suggested the useful terms *principal* and *accessory* shocks, though his division of the latter into *preparatory* and *consecutive* shocks has met with less general approval.

Though he continued to serve on the Commission, Forel's active share in its work seems to have ended in 1888. The Commission was, however, joined by other recruits, for instance, by J. Früh of Zurich, afterwards its president, A. de Quervain of Zurich, later secretary, etc., until the total number of members had risen in 1906 to fifteen. In 1891, the annual reports were published for the first time in the *Annalen* of the Swiss Meteorological Office, and, in this journal, all the later reports, as well as those of its successor, have appeared.

After an existence of rather more than thirty years, the work of the Seismological Commission was taken over by the Meteorological Office, and, since 1913, the annual reports have been those of the Swiss Earthquake Service, A. de Quervain continuing to act as its secretary.

Throughout the whole time, the work of both bodies has been almost confined to the study of earthquakes felt in Switzerland. How valuable this study has been may be gathered from a summary, given by its president J. Früh, of the Commission's work for thirty years*. From this, we learn that, during the years 1880–1909, the total number of shocks recorded was 998, including 195 principal earthquakes that originated within the area of Switzerland. Of the former, 377 occurred in winter (Dec.–Feb.), 231 in spring, 155 in summer, and 235 in autumn; while 729 were felt from 8 p.m. to 8 a.m. and 269 from 8 a.m. to 8 p.m. During the eleven years 1913–23, the total number of shocks recorded by the Earthquake Service was 531, 217 being in winter, 113 in spring, 94 in summer and 107 in autumn.

142. Austrian Earthquake Commission. The destructive earthquake of Laibach (art. 136) occurred on 14 Apr. 1895. A few days later, on 25 Apr., a special commission for the study of Austrian earthquakes was instituted by the Vienna Akademie

* *Schweiz. Natf. Ges. Verh.* 1911, 24 pp.

der Wissenschaften. Three aims were set before the Commission: (i) the compilation of a catalogue of the earthquakes that have disturbed the eastern Alpine district; (ii) the erection of registering instruments at certain selected stations; and (iii) the formation of a widespread system of observers.

The formation of the earthquake-catalogue was entrusted to Rudolf Hoernes (art. 134), but, apparently, the task was left unfinished at his death in 1912. Seismographs were erected at several observatories, such as Vienna, Trieste, Kremsmünster and Lemberg. The system of observers, as with the Swiss Seismological Commission, was based on the provinces. For each of fifteen districts, a reporter was appointed, whose duty it was to collect the accounts of the observers and forward them to the central office. The Commission was fortunate in its reporters, Hoernes being responsible for Styria, Ferdinand Seidl for Görz and Carniola, Eduard Mazelle (1862–) for the Trieste district, and Friedrich Becke (1855–) and Johann Woldrich for Bohemia. By the end of 1896, the total number of observers enlisted amounted to 1751*.

143. During the first four years (1897–1900), the reports of the Commission were published in the *Sitzungsberichte* of the Akademie (Math.-Naturw. Classe). With the beginning of the present century, they were issued as separate pamphlets under the title *Mittheilungen der Erdbeben-Kommission*, etc. The reports were written by individual members, never as the joint work of the Commission. The first series contained 21 reports, the second (up to 1913) 48, the whole occupying about 4000 pages†.

It is a matter for regret that these valuable reports should be so little known. As most of them are the work of men still living, their consideration lies outside our range. I must confine myself to touching briefly on their contents. We have, in the first place, the annual lists of earthquakes for the years 1896–1903 by Edmund von Mojsisovics (1839–1907)‡. Of special

* Wien, *Ak. Sber.* vol. 106, pt. 1, 1897, pp. 20–45.

† Since the beginning of the War, thirteen numbers of the *Mittheilungen* have been published.

‡ Ser. 1, nos. 1, 5, 10, 18; ser. 2, nos. 2, 10, 19, 25.

earthquakes, many investigations were made, such as those of the Scheibbser earthquake of 1876, the Murztaler earthquake of 1885, the Graslitzer earthquake of 1897, the Sinj earthquake of 1898, the Styrian earthquakes of 1898 and 1904, the Salonica earthquake of 1902 and the Macedonian earthquake of 1904*. With these, may be included the studies on brontides by J. Knett and J. N. Woldrich†. The seismograph records at the four observatories mentioned above occupy a score of reports‡. Lastly, we have the valuable theoretical inquiries by H. Benndorf (1870–), W. Lâska and W. Trabert (1863–) on the light thrown by the velocity of the earthquake-waves on the nature of the earth's interior§.

CONCLUSION

144. The principal results achieved by workers in Germany, Austria and Switzerland from 1845 to 1900 are the following: (i) the invention of some useful terms such as centre and epicentre, pleistoseist and homoseist, principal shock and accessory shocks, fore-shocks and after-shocks (art. 119); (ii) the study of some important earthquakes, and especially those in Rhineland in 1846, the Visp valley in 1855, Sillein in 1858, Central Germany in 1872, Belluno in 1873, Herzogenrath in 1873 and 1877, Agram in 1880, and Laibach in 1895 (arts. 120, 121, 123–126, 131, 132, 135, 136); (iii) the construction of homoseists and the use made of time-records in determining the depth of the focus (arts. 123–125, 132); (iv) the construction of isoseismal lines, especially of those depending on various degrees of damage and on the ultimate perception of the shock, and the determination of the epicentre by their means (art. 120); (v) the realisation of the connexion between earthquakes and definite structural lines in the earth's crust and the recognition of the possible multiple origin of some earthquakes (arts. 132–134); and (vi) the foundation of the Swiss Seismological Commission in 1878 and of the Austrian Earthquake Commission in 1895 (arts. 138–143).

* Ser. 1, nos. 8, 13; ser. 2, nos. 13, 17, 24, 27, 32, 40.

† Ser. 1, nos. 9, 20, 21.

‡ Vienna, ser. 2, nos. 33, 35, 39, 41, 45, 47; Trieste, ser. 1, nos. 11, 17; ser. 2, nos. 5, 11, 20, 30; Kremsmünster, ser. 1, no. 15; ser. 2, nos. 4, 12, 21, 26; Lemberg, ser. 2, nos. 1, 9, 22, 28.

§ Ser. 2, nos. 14, 23, 29, 31, 37, 38, 46.

CHAPTER VIII

THE STUDY OF EARTHQUAKES IN THE UNITED STATES

145. The United States possess three or four seismic districts of great interest and others of less importance, and these from time to time have gained the attention of painstaking observers. To the former class belongs the State of California, in which earthquakes are frequent and sometimes of destructive violence. One of their earliest students was John B. Trask (*c.* 1823–79), a skilful and original physician, who attended to them from 1850 to 1858 and, during these nine years, recorded altogether 98 earthquakes*. If his lists are now known to be far from complete, and if, since his time, some of the gaps which he left have been partially filled in, he must none the less be regarded as a pioneer of Californian seismology.

146. Some of the Californian earthquakes have been restudied in later years. Among these is the Owen's Valley earthquake of 26 Mar. 1872, first investigated by Josiah Dwight Whitney (1819–96), professor of geology at Harvard university. Whitney's report, unfortunately, was not published in a scientific journal, and it is only lately that its principal contents have become generally known†. Many of his observations are of permanent interest, especially those on the structure of the valley, the sound-phenomena of the earthquake, the landslips, the water-wave on Owen's Lake, the earthquake lakes and pools, and the remarkable fault-movements, the scarps of which at the time of his visit, two months after the earthquake, had lost none of their freshness.

Nor were they much weather-worn when, eleven years later, they were examined by Grove Karl Gilbert (1843–1918), of the

* *Amer. Jl. Sci.* vol. 22, 1856, pp. 110–116; vol. 23, 1857, pp. 341–346; vol. 25, 1858, pp. 146–148; vol. 26, 1858, pp. 296–298; vol. 28, 1859, pp. 447–448.

† *Overland Monthly*, vol. 9, 1872, pp. 130–140, 266–278. A summary of these papers was given by C. G. Rockwood (1843–1913) in *Amer. Jl. Sci.* vol. 4, 1872, pp. 316–318, and long extracts have been usefully reprinted by W. H. Hobbs (1864–) in *Beitr. Geophys.* vol. 10, 1910, pp. 356, 364–372, 379.

U.S. Geological Survey, well-known by his monographs on the Henry Mountains, Lake Bonneville, etc., and later by his help in the investigation of the Californian earthquake of 1906. Gilbert's brief paper on the earthquakes of the Great Basin*, in which the scarps are described, is notable for a clear account of the theory which refers earthquakes to the friction called into action during the growth of faults†.

147. On the opposite side of the continent, the earthquakes of New England found a historian in William T. Brigham (1841–*c.* 1923), who had already (1866) written an important memoir on the volcanoes of the Hawaiian Islands. His catalogue of New England earthquakes‡ contains 231 entries from 1638 to 1869, the Pilgrim Fathers having landed in 1620. It is evidently the result of very extensive reading§ and includes several interesting accounts, especially that of the remarkable series of earth-sounds heard at East Haddam (Connecticut). He defines three seismic regions in the States considered—one near Montreal, the second around the mouth of the Merrimac River, and the third including New Haven, Lyme and East Haddam.

148. The earliest student of earthquakes in general in the United States was Charles Greene Rockwood (1843–1913), who was educated at Yale university, and served as professor of mathematics at Bowdoin College, Maine (1868–73), Rutgers College, N.J. (1874–77), and Princeton university, N.J. (1877–1905). He died so many years after his latest contribution to science (1887) that he passed away almost forgotten by the next generation‖. So far as can be judged from his printed papers, he

* *Amer. Jl. Sci.* vol. 27, 1884, pp. 49–53; also *Lake Bonneville, U. S. Geol. Surv. Mon.* 1890, p. 361.

† In 1907, the fault-scarps were still clear enough to be photographed and mapped by Willard D. Johnson of the U.S. Geological Survey. Both maps and photographs are reproduced by W. H. Hobbs in the paper referred to above (pp. 352–384).

‡ *Boston Soc. Nat. Hist. Mem.* vol. 2, 1871, pp. 1–28.

§ In the same volume (pp. 241–247), Albert Benoit Marie Lancaster (1849–), partly from his own inquiries, chiefly from Perrey's memoir on the earthquakes of the United States and Canada, adds 52 earthquakes to Brigham's list.

‖ No biographical notice worthy of the name seems to have appeared in any scientific journal. For the above facts, I am indebted to Mr V. Lansing Collins, secretary of Princeton university.

seems to have been an interested, but perhaps not a very energetic, student of earthquakes, and to have welcomed the more valuable contributions of his successors without the faintest trace of jealousy.

With two exceptions, Rockwood's scientific papers deal entirely with seismic and volcanic phenomena. His first paper (1871) is an interesting study of the daily motion of a brick tower caused by the sun's heat, and the last (1887) describes an insect fight. His writings on earthquakes (excluding popular papers) fall into two groups, (i) a series of 16 papers (1872–86) on American earthquakes, and (ii) several summaries of recent work, the more important being those published in the Reports of the Smithsonian Institution for 1884, 1885 and 1887. Though useful from an educational point of view, the second group of course contains no fresh contributions to seismology.

The first group* begins with a brief account of the New England earthquake of 9 Jan. 1872, and is followed by four papers on recent earthquakes and eleven on recent American earthquakes. The total number of United States earthquakes in the various lists is 426, excluding 34 doubtful shocks. In the first fourteen papers, the descriptions are invariably brief. They contain little more than the date of the earthquake and the area chiefly affected. It is, unfortunately, only as the series is drawing to a close that the accounts enter into detail, and it is worthy of notice that their increased value coincides with the adoption of a simple scale of intensity containing six degrees†. This scale rendered possible the construction of isoseismal lines, and, for the first time, Rockwood illustrates his descriptions by maps containing such lines, three intensity curves, as he calls them, being given for the New York earthquake of 10 Aug. 1884 and

* *Amer. Jl. Sci.* vol. 3, 1872, pp. 233–234; vol. 4, 1872, pp. 1–4; vol. 5, 1873, pp. 26–263; vol. 6, 1873, pp. 40–44; vol. 7, 1874, pp. 384–387; vol. 9, 1875, pp. 331–334; vol. 12, 1876, pp. 25–30; vol. 15, 1878, pp. 21–27; vol. 17, 1879, pp. 158–162; vol. 19, 1880, pp. 295–299; vol. 21, 1881, pp. 198–202; vol. 23, 1882, pp. 257–261; vol. 25, 1883, pp. 353–360; vol. 27, 1884, pp. 358–364; vol. 29, 1885, pp. 425–437; vol. 32, 1886, pp. 7–19.

† The scale at first consisted of adjectives only, namely: (1) very slight (2, 3), (2) light (4), (3) moderate (5, 6), (4) strong (6, 7), (5) severe (8), (6) destructive (9, 10); the corresponding degrees (in brackets) of the Rossi-Forel scale being added in 1886.

two for the Ohio earthquake of 19 Sept. 1884. In the final paper (1886) appears a sketch-map of California on which are depicted the outlines of the disturbed areas of eleven earthquakes originating in that State during the year 1885. His correlation of two of the earthquakes of 1884 with the geological structure of the epicentral regions makes one regret that his series of notes on American earthquakes should have come to an end just as they began to show promise of widening interest.

149. During the years 1876–77, some interesting experiments on the velocity of earth-waves were made by Henry Larcom Abbot (1831–), a general in the U. S. Army. The more important were those resulting from the explosion of 50,000 lb. of dynamite at Hallet's Point in New York harbour. Others were also made in which the charges were 400 and 200 lb. of dynamite and 70 lb. of gunpowder. Altogether, Abbot's observations were made on six explosions and along twelve lines of various lengths (1·169 to 12·769 miles). The times of the explosion and of the arrival of the first tremors were in each case electrically recorded on the same moving sheet of paper, the latter instant being usually determined by two observers watching the ruffled surface of a mercury bath with telescopes magnifying six and twelve times. From these experiments, Abbot drew the following conclusions: (i) The mean of six observations with the lower power gave a velocity of 4225 ft. per sec., and the mean of the same number with the higher power gave one of 7475 ft. per sec. (ii) There is some, though not very decisive, evidence to show that the velocity increases with the intensity of the disturbance, the velocities given by the explosion of 400 lb. of dynamite, 200 lb. of dynamite, and 70 lb. of gunpowder, being respectively 8814, 8730 and 8415 ft. per sec. (iii) The velocity seems to decrease as the wave advances, being, with a charge of 200 lb. of dynamite, 8730 ft. per sec. for one mile and 5280 ft. per sec. for five miles; and, with a charge of 50,000 lb. of dynamite, 8300 ft. per sec. for about eight miles and 5300 ft. per sec. for about thirteen miles*.

* *Amer. Jl. Sci.* vol. 15, 1878, pp. 178–184. Mallet criticised Abbot's results in two papers (*Phil. Mag.* vol. 4, 1877, pp. 298–302; vol. 5, 1878, pp. 358–362), noticing especially the remarkable inequalities in Abbot's estimates.

150. Seismology in America is indebted very largely to the labours of students in other fields of science. In California, the most unstable district in the United States, the work begun by Rockwood was continued for a time by Edward Singleton Holden (1846–1914), who was director of the Lick Observatory on Mount Hamilton from 1885 to 1898. In 1887 Holden installed a set of Ewing seismographs in this observatory; a second set was placed in the university of Berkeley, while seven other stations in California and one in Nevada were provided with duplex pendulum seismographs of the Ewing type. Holden's chief work in seismology was the compilation of a *List of recorded earthquakes in California, Lower California, Oregon and Washington Territory**. Based mainly on the catalogues of Mallet, Perrey, Rockwood, Fuchs and Trask, but to a great extent also on manuscript notes, this list contains entries of 948 earthquakes between the years 1769 and 1887 inclusive. The accounts in but few cases give more than the date, the place or places chiefly affected, and (when known) the intensity according to the Rossi-Forel scale. Brief as they are, however, they furnish the groundwork for a fuller history, for the detailed mapping of the zones of one of the most interesting seismic districts in the world.

This list was followed by an attempt to estimate the total earthquake-intensity at San Francisco from 1808 to 1888†, the chief interest of which now lies in a side issue. Making use of the Japanese seismograph records obtained by Ewing, Milne and Sekiya, Holden computed the maximum acceleration corresponding to each of the degrees 1 to 9 of the Rossi-Forel scale (art. 101). In so doing, he accomplished more than he thought, for he had in reality devised the first absolute scale of seismic intensity.

It was evidently Holden's intention that his catalogue of Californian earthquakes should be followed by annual lists. That for the year 1888‡ he prepared himself. The lists for the next ten years were published as *Bulletins* by the U.S. Geological Survey. That for the year 1889 (Bull. no. 68) was written by James

* Sacramento, 1887, 78 pp.
† *Amer. Jl. Sci.* vol. 35, 1888, pp. 427–431.
‡ *Amer. Jl. Sci.* vol. 37, 1889, pp. 392–402.

Edward Keeler (1857–1900), then assistant, afterwards (1898) director, of the Lick Observatory. The lists for 1890 and 1891 (Bull. no. 95) were compiled by Holden, and those for the remaining years (Bull. nos. 112, 114, 129, 147, 155 and 161) by Charles Dillon Perrine (1867–), afterwards (1909) director of the Argentine National Observatory at Cordoba and the discoverer of the sixth and seventh satellites of Jupiter. After 1898, when Holden had retired from the Lick Observatory, the issue of annual lists was discontinued.

THE CHARLESTON EARTHQUAKE OF 31 AUGUST 1886

151. Among the earthquakes of the United States, two, that occurred near large towns, stand out in special prominence—the Charleston earthquake of 1886 and the Californian earthquake of 1906—and it is mainly to them that the interest in seismology now so widespread in that country must be ascribed. Two others —the New Madrid earthquakes of 1811–12 and the Alaskan earthquakes of 1899—that visited less populous districts, have also been studied after the lapse of some years.

The Charleston earthquake occurred on 31 Aug. 1886 in a district which, up to that date, might have been regarded as almost aseismic. Early in September, W. J. McGee (1853–1912), of the U.S. Geological Survey, and Thomas Corwin Mendenhall (1841–1924)*, of the U.S. Signal Service, proceeded to the central area. Another careful observer, Earle Sloan, spent two months in a detailed examination of the epicentral tracts. At the same time, Edward Everett Hayden (1858–), of the U.S. Navy, collected records of the shock from all parts of the disturbed area, special attention being paid to observations of the time. The various accounts thus collected, nearly 4000 in number from about 1600 stations, were placed in the hands of Captain (afterwards Major) C. E. Dutton (1841–1912), of the U.S. Geological Survey.

152. Clarence Edward Dutton was born on 15 May 1841. Educated at Yale College, where he graduated in 1860, he was

* Mendenhall's report was published in the *Monthly Review* (U.S. Signal Service) for Aug. 1886, and was reprinted in *Nature*, vol. 35, 1886, pp. 31–33.

on active service in the civil war from the autumn of 1862 until its close in 1865. In 1875 he joined the U.S. Geological Survey, and the next ten years were spent mainly in the study of the great plateau region of the West, and in the preparation of his valuable monographs on the Grand Cañons of Colorado and the High Plateaus of Utah. In 1882, he visited the Hawaiian Islands, and it may have been due to his work on this volcanic district that he was selected as the official investigator of the Charleston earthquake. He retired from the army in 1901 and died on 4 Jan. 1912*.

Not many seismologists have made so few original contributions to the science. Two or three minor papers† on the Charleston earthquake were afterwards incorporated in the complete report‡—one that will always hold an honourable place among earthquake monographs—and in 1904 he published a popular work on earthquakes§, clearly and brightly written, but not quite representative of our knowledge at the beginning of the century. His single contribution to seismology is thus his report on the Charleston earthquake.

153. In this report are to be found many interesting notes on the occurrence of visible surface-waves in the epicentral tracts, the formation of sand-craters, and the variation in the intensity of the shock throughout the disturbed area. Such descriptions are, however, common to the accounts of all great earthquakes. The outstanding results of the investigation fall under three headings—(i) the determination of the double epicentre (pp. 270–310), (ii) the attempt to ascertain the depths of the corresponding foci (pp. 311–320), and (iii) the estimate of the mean velocity of the earth-waves (pp. 355–389).

(i) The observations under the first heading are of unusual interest, the Charleston earthquake being one of the earliest in which the existence of two detached epicentres was recognised (art. 132). Their discovery was by no means a simple matter, for

* *Amer. Seis. Soc. Bull.* vol. 1, 1911, pp. 137–142; *Amer. Jl. Sci.* vol. 33, 1912, pp. 387–388.

† *Science*, vol. 10, 1887, pp. 10–11, 35–36; *Washington Phil. Soc. Bull.* vol. 10, 1888, pp. 17 (bis)–19 (bis); *Nature*, vol. 36, 1887, pp. 269–273, 297–303.

‡ The Charleston Earthquake of 31 August, 1886. *U.S. Geol. Surv. Rep.* 1887–88 (1889), pp. 203–528.

§ *Earthquakes in the Light of the New Seismology*, Putnam, 1904.

the epicentral district, which lies from 10 to 20 miles to the west and north-west of Charleston, is in some places barren and forest-clad, in others swampy, and in all sparsely inhabited. That there were two epicentral tracts is clear from the increased frequency of sand-craters and fissures within them, from the injury to the three lines of railway that cross the district from Charleston, and also from the nature of the shock. The boundaries of the tracts are naturally ill-defined, but the large one near Woodstock station is approximately a circle 20 miles in diameter, and the smaller one near Rantowles station an ellipse with axes 9 and 6 miles in length. The distance between their centres is about 13 or 14 miles.

(ii) Less successful than the definition of the epicentral tracts was the attempt to determine the depths of the corresponding foci. Even if the method used were irreproachable, its application is fraught with uncertainty. Assuming that the intensity at any point varies inversely as the square of the distance from the focus, Dutton shows that the rate at which the intensity declines must be a maximum at a distance from the epicentre which bears to the depth of the focus the ratio 1 to $\sqrt{3}$. For the Woodstock epicentre, he found this distance to be nearly 7 miles, and, for the Rantowles epicentre, about $4\frac{1}{2}$ miles. The corresponding depths of the foci would thus be about 12 and 8 miles*.

(iii) A very important result of Dutton's inquiry was his estimate of the mean velocity of the earth-waves, probably the most accurate up to that time made for any earthquake. In this, he was favoured by two conditions—the existence of the standard time-system, by which the exact time is transmitted once a day to every telegraph-office in the country, and the great distances traversed by the earth-waves, 60 good records coming from

* Dutton's method has been applied in six other earthquakes, the estimated depths being 21 miles for the Constantinople earthquake of 1894, $4\frac{1}{2}$ miles for the Syracuse earthquake of 1895, $4\frac{1}{3}$ miles for the Marsica (Italy) earthquake of 1904, between 21 and 40 miles for the Kangra earthquake of 1905, 6 miles for the Marsica earthquake of 1915, and 8 or 9 miles for the Srimangel (Assam) earthquake of 1918 (*Ann. Géogr.* 1895, p. 164; Roma, *Uff. Centr. Meteorol. Ann.* vol. 16, pt. 1, 1894, p. 7; *Ital. Soc. Sism. Boll.* vol. 18, 1914, p. 484; *India Geol. Surv. Mem.* vol. 38, 1910, pp. 332–334; *Ital. Soc. Sism. Boll.* vol. 19, 1915, p. 193; *India Geol. Surv. Mem.* vol. 46, 1920, p. 47).

places more than 500 miles from the Woodstock epicentre. The total number of time-records at his disposal was 316, of which 130 were rejected for various reasons and 45 were obtained from stopped clocks. The remaining observations were divided into three classes, the first containing 5 records in which the epoch of the first maximum was given to within 15 seconds, the second 11 records to the nearest minute or half-minute, and the third 125 records of more uncertain value. The mean velocities obtained from the three classes were respectively 5·21, 5·19 and 4·85 km. per sec. In estimating the mean velocity from all the observations, the weights of the three groups were taken as inversely proportional to the squares of the probable errors, or as 2 : 1 : 4, the resulting value being 5·18 km. per sec. It is interesting to notice how closely this estimate accords with that obtained from the Oppau explosion of 21 Sept. 1921. This was registered at five European observatories at distances from Oppau ranging from 68 to 227 miles, the recorded times giving a velocity of 5·4 km. per sec. for the first preliminary waves and of 3·15 km. per sec. for the second. Such values refer of course to the superficial sedimentary layer, the values for the upper layer of the underlying crust being considerably higher (7·1 and 4·0 km. per sec.)*. Thus, they confirm Dutton's suggestion of the moderate depth of the foci.

THE CALIFORNIAN EARTHQUAKE OF 18 APRIL 1906

154. Just as the Yokohama earthquake of 1880 led to the founding of the Seismological Society of Japan (art. 187), the Mino-Owari earthquake of 1891 to that of the Imperial Earthquake Investigation Commission (art. 214), and the Laibach earthquake of 1895 to that of the Austrian Earthquake Commission (art. 142), so the Californian earthquake of 1906 marks another epoch in seismology. It was followed by two noteworthy results—(i) the preparation and publication of the elaborate report on the earthquake, and (ii) the foundation of the Seismological Society of America. If I were to keep strictly to the lines marked out in the first chapter, the reference to both results

* D. Wrinch and H. Jeffreys, *Astr. Soc. M. Not., Geoph. Sect.* vol. 1, 1923, pp. 15–22.

would be slight or altogether absent, as most of the workers are still with us. This chapter, however, would be incomplete without a few words on the principal achievements of American seismologists.

155. Report on the Californian Earthquake of 1906. This great earthquake occurred on 18 April. Three days later, the State Earthquake Investigation Commission was nominated by the Governor of California. It consisted of four geologists and four astronomers. The main work of the inquiry naturally fell to the geologists—Andrew Cowper Lawson (1861–), professor of geology in the university of California, G. K. Gilbert (1843–1918; art. 146), Harry Fielding Reid (1859–), professor of geology in Johns Hopkins university, Baltimore, and John Casper Branner (1850–1922; art. 146), professor of geology in Stanford university, California. A. C. Lawson acted as chairman of the Commission and edited the first volume of the report, the whole of the second volume being written by H. F. Reid. The astronomical members were Arnim Otto Leuschner (1868–), the secretary to the Commission, George Davidson (1825–1911), professor of astronomy in the university of California, Charles Burckhalter (1849–1923), and William Walter Campbell (1862–), the director of the Lick Observatory. They were assisted by many colleagues, who helped to trace the course of the San Andreas rift, to measure the displacements on either side, and to map the variations of intensity throughout the disturbed area. But it is worthy of notice that, with the exception of G. K. Gilbert, none of them was at that time known to seismologists as a student of earthquakes. There is thus in the earlier volume an absence of reference to corresponding features in other earthquakes. Indeed, if the Californian earthquake were the only shock known to mankind, the attention paid to it could hardly have been more exclusive. On the other hand, the investigation has gained in freshness and in freedom from the trammels of old methods.

No report on any previous earthquake* has been issued on so

* *The Californian Earthquake of April 18, 1906: Report of the State Investigation Commission*, 2 vols. and atlas; vol. 1, 1908, 451 pp. and 146 plates; vol. 2, 1910, 192 pp. and 2 plates. Published by the Carnegie Institution of Washington.

liberal a scale. It is, for instance, about twice the length of Mallet's great report on the Neapolitan earthquake of 1857 (art. 73). Nor has any been so finely illustrated. In addition to a folio atlas of 40 plates, there are more than three times that number inserted in the two quarto volumes.

The account of the earthquake given in this report is the joint work of a large number of observers. Whenever possible, they have been allowed to speak for themselves in short notes and papers, so neatly worked into the text that, in reading it, there seems to be no breach of continuity. The advantages of this mode of collaboration are nowhere better illustrated than in the description of the San Andreas rift. Traced by different observers for a distance of at least 190 miles, except for a few short submarine interruptions, there can be little doubt that it reappears after a somewhat longer break farther to the north, so that the total length of the displaced portion must be about 270 miles. On the whole, the path of the fault is a slightly curved line running in a general north-west and south-east direction, and, to the north of San Francisco, keeping close to the Californian coast. Its mere length is not, however, its most remarkable feature. Throughout its whole extent, the crust on each side was displaced, that on the south-west side of the fault to the north-west, and on the north-east side to the south-east. The amount of the horizontal displacement varies considerably. As a rule, it lies between 8 and 15 feet, and in no place is it known to exceed 21 feet. In other earthquakes, there have been more pronounced changes of level along the fault, but, in no other, have the observed horizontal movements been so persistent over so vast a length. In none, certainly, have the fault-movements been so carefully studied or the investigations of the geologist been aided so effectively by the researches of the biologist or the measurements of the surveyor.

Of the special memoirs inserted in the text, the most important perhaps are those by J. F. Hayford (1868–1925) and A. L. Baldwin on the geodetic measurements of the earth-movements (vol. 1, pp. 114–145), B. A. Baird on the relative positions of the monuments erected on either side of the fault in order to detect and measure future movements (vol. 1, pp. 152–159), and H. O.

Wood on the distribution of intensity in San Francisco (vol. 1, pp. 220–254). The second volume, by H. F. Reid, consists of valuable discussions on the mechanics of the earthquake, and the permanent displacements of the ground, and on the instrumental records of the earthquake and the propagation of the earthquake-waves.

156. The Seismological Society of America. It was only natural that an earthquake so strong and so destructive should rouse an abiding interest in earthquakes, that it should lead to the formation of a seismological society. A preliminary meeting for this purpose was held in a temporary frame building. It was called by Alexander George McAdie (1863–), then professor of meteorology, U.S. Weather Bureau. The next meeting, at which the constitution of the society was fixed, was held on 20 Nov. 1906, and it was provided that a board of twelve directors should be elected annually in April and that they should choose their own officers who should also be officers of the society. The first president was George Davidson, of English birth, but educated in the United States, for several years engaged in the triangulation of the Pacific coast, and from 1873 professor of astronomy in the university of California. After a little more than two years' service, he resigned owing to failing health, and was followed by A. C. Lawson, J. C. Branner, A. G. McAdie, Otto Klotz (1852–1923), Bailey Willis (1857–) and others. The secretary throughout has been Sidney Dean Townley (1867–), professor of applied mathematics in Stanford university.

157. Of the various presidents, John Casper Branner must be regarded as a builder, rather than as a founder, of the Society, for he took no part in the early meetings. But to no one member does the Society owe so deep a debt. Of sterling honesty and public spirit, contributing freely of his time and money*, and unthinking of his own credit, his efforts trebled the number of members, he founded, and to some extent supported, the *Bulletin*, and in 1919 he presented Montessus' almost unique

* From 1911 to 1921, at his own expense, he provided investigators of seven earthquakes in different parts of California.

library of several thousand volumes and pamphlets on earthquakes to Stanford university*.

158. Though the Society was founded at the close of 1906, it was not until March 1911 that the directors felt justified in issuing the first part of a quarterly *Bulletin*. It is interesting to notice the gradual expansion of this journal and the increasing importance of its articles. The financial difficulties that faced it at times during the first thirteen years of its existence have been removed by the support recently (1924) given to it by the Carnegie Institution of Washington. I cannot do more than refer briefly to some of the more important memoirs. Several recent earthquakes on the western coast have been investigated, for instance, the Imperial Valley (Cal.) earthquake of 22 June 1915 by Carl H. Beal and the Pleasant Valley (Nev.) earthquake of 2 Oct. 1915 by J. Claude Jones (1877–)†. Some important studies have been made of the earthquakes of other lands, such as those by Noah Fields Drake (1864–) on destructive earthquakes in China, J. C. Branner on earthquakes in Brazil, J. Scherer on great earthquakes in the island of Haiti, M. Saderra Masò and W. D. Smith (1880–) on earthquakes in the Philippines, H. O. Wood on the Hawaiian earthquakes of 1868, and Stephen Taber (1882–) on earthquakes in the Atlantic coastal plain, the southern Appalachians, and the greater Antilles‡. Among memoirs on general subjects may be mentioned that of Leo A. Cotton on earthquake frequency with special reference to tidal stresses in the lithosphere§. But perhaps the most important work so far of the Seismological Society has been the publication of the great map of earthquake-faults in California||.

THE NEW MADRID EARTHQUAKES OF 1811–12 AND THE ALASKAN EARTHQUAKES OF 1899

159. New Madrid Earthquakes of 1811–12. When Lyell visited the central area of these earthquakes in 1846, their

* S. D. Townley, *Amer. Seis. Soc. Bull.* vol. 12, 1922, pp. 1–11.

† *Amer. Seis. Soc. Bull.* vol. 5, 1915, pp. 130–149, 190–205.

‡ Vol. 2, 1912, pp. 40–91, 105–117, 124–133, 161–180; vol. 3, 1913, pp. 151–186; vol. 4, 1914, pp. 108–160, 169–203; vol. 6, 1916, pp. 218–226, and vol. 12, 1922, pp. 199–219.

§ Vol. 12, 1922, pp. 49–198.

|| Vol. 13, 1923.

remarkable effects were still plainly visible (art. 43). Even after the lapse of nearly a century, they were distinct enough to be the subject of a detailed study by Myron Leslie Fuller (1873–) of the U.S. Geological Survey*. Especially marked are the ejections of sand and the long narrow domes and sunk-lands. (i) Sand-craters are formed in nearly all great earthquakes, through the violent ejection of water and sand from fissures in the ground. Fuller describes two other forms—sand-blows and sand-sloughs—which resulted from the quiet extrusion of sand. The former are low, usually circular, mounds of all diameters up to 100 feet and more. Sand-sloughs are low ridges alternating with shallow troughs in which water is collected in long narrow pools. Both are probably caused by the unequal settling of the alluvial deposits into the water-bearing stratum below. (ii) Of still greater interest is the warping of the surface in domes and sunk-lands. The three principal domes, the largest of which is about 15 miles long and 20 feet high, form a long narrow ridge. To the east of it lies a trough occupied by the Mississippi and the Reelfoot Lake; to the west another trough and, still farther, a second ridge and third trough parallel to the others. The cause of the warping is not quite clear, but it may be due to further sinking of the basin. This, at any rate, would account for the preponderance of subsidence over uplift.

160. Alaskan Earthquake of 1899. More striking even than the effects of the New Madrid earthquakes were those of the great Alaskan earthquakes of 3 and 10 Sept. 1899. They were studied in 1905 and 1906 by Ralph Stockman Tarr (1864–1912), professor of physical geography at Cornell university, and Lawrence Martin (1880–). Though their admirable report on the earthquakes† is full of interest, two points only can be referred to here. (i) In no other earthquakes known to us have such great changes of elevation occurred. In this case, the measurements are absolute, for they are all referred to sea-level. They were determined partly by the rise of rock-benches cut by the sea in the cliffs, but mostly by the height of dead barnacles still

* *U.S. Geol. Surv. Bull.* no. 494, 1912, 112 pp.
† *U.S. Geol. Surv. Prof. Paper*, no. 69, 1912, 135 pp.

adhering to the rock. The greatest uplift known to us, of 47 ft. 4 in. occurred on the west coast of Disenchantment Bay. A very remarkable feature is the rapidity with which the elevation changed. At one point on the coast referred to, the uplift was 42 feet, about a mile to the west it was 30 feet, and about a quarter of a mile farther on it had dropped to 9 feet. (ii) A few miles to the north of Yakutat Bay lie the lofty St Elias and other ranges, from which many glaciers descend to the coast. From these mountains, great snow-avalanches fell and accumulated in the glacier reservoirs. In 1899, all the Yakutat Bay glaciers were retreating. Then, one by one, they began to advance, the shortest some time before 1905, the longest being still unaffected in 1910. In all of them, the advance was abrupt; the surface, previously smooth and easily crossed was transformed into a wilderness of pinnacles and crevasses; the glaciers thickened at their lower ends; and finally, after advancing several hundred yards in ten months or less, the effects of the increased supply of snow were spent, and the glaciers rapidly returned to a stagnant condition.

CONCLUSION

161. The admirable work of seismologists in the United States has led to results principally on the following lines: (i) the investigation of special earthquakes, above all of the New Madrid earthquakes of 1811–12, the Owen's Valley earthquake of 1872, the Charleston earthquake of 1886, the Alaskan earthquakes of 1899, and the Californian earthquake of 1906 (arts. 146, 151–153, 155, 159, 160); (ii) the compilation of earthquake-catalogues for special districts like California and New England (arts. 145, 147); (iii) the estimate of the velocity of earth-waves in the Charleston earthquake of 1886 (art. 153); (iv) the study and accurate measurements of the displacement of the earth's crust in the New Madrid, Owen's Valley, Alaskan and Californian earthquakes (arts. 155, 159, 160); and (v) the foundation in 1906 of the Seismological Society of America (arts. 156–158).

CHAPTER IX

FERNAND DE MONTESSUS DE BALLORE

162. During the interval of about twenty years that elapsed between Perrey's retirement from Dijon (1867) and Montessus' first important memoir (1888), scientific men in France almost disregarded earthquakes. For a few years longer Perrey compiled the annual lists of earthquakes from his retreat at Lorient. Antoine d'Abbadie (1810–97) had built his nadirane* at Abbadia near Hendaye in 1863, in the hope of measuring deviations of the vertical, and Charles Wolf (1827–1918) erected a somewhat similar instrument† in 1883 in the Paris observatory. Besides d'Abbadie and Wolf, whose interests were mainly astronomical, the chief contributor to seismology was Ferdinand André Fouqué (1828–1904), professor of geology in the Collège de France. In 1861, he accompanied Sainte-Claire Deville (1818–81) on an expedition to Vesuvius. He also helped to investigate the eruption of Etna in 1865, Santorin in 1866, and Terceira (Azores) in 1867. In addition to his task as editor of the report on the Andalusian earthquake of 1884, Fouqué wrote a popular textbook on earthquakes‡, in the second part of which (pp. 248–326), he followed Lyell's example (art. 43) in giving brief accounts of the Ilopango earthquake of 1854, the great Japanese earthquake of the same year, the Ischian earthquake of 1883, the Andalusian earthquake of 1884, and the Riviera earthquake of 1887.

163. The Andalusian earthquake occurred on 25 Dec. 1884. It was investigated by three commissions—The Spanish under M. F. de Castro§, the Italian, consisting of Taramelli and Mercalli (art. 106), and the French, appointed by the Academy

* Paris, *Ac. Sci. C. R.* vol. 34, 1853, pp. 942–943; vol. 61, 1865, p. 838; vol. 89, 1879, pp. 1016–1017; vol. 98, 1884, p. 322–323.

† Paris, *Ac. Sci. C. R.* vol. 97, 1883, pp. 229–234.

‡ *Les Tremblements de Terre*, 1888, 328 pp.

§ *Terremotos de Andalucía: Informe de la comisión nombrada para su estudio*, etc., Madrid, 1885, 107 pp.

of Sciences. Of the latter, Fouqué was director, and he was assisted by A. Michel Lévy (1845–1911), Marcel Bertrand (1847–1907), Charles Barrois (1857–), Wilfrid Kilian (1862–1925), and others. It was thus a corps of geologists, none of whom, with the exception of Kilian, who had written a brief note on an earthquake at Grenoble in 1883, had previously studied any earthquake. It is therefore not surprising that the commission should have been more interested in the geology of the central district than in the earthquake itself. Of their long report*, only one-tenth deals with the earthquake and the velocity of earthquake-waves. The remainder consists of five memoirs on the geology and paleontology of the district.

To the chapter on the earthquake-phenomena (pp. 9–55) all the members of the commission seem to have contributed. The map of the central region shows three isoseismal lines, the innermost bounding the area of destruction, and the next that in which houses were damaged. The earthquake was the first in which magnetic instruments were widely disturbed—at Lisbon, Parc Saint-Maur (Paris), Greenwich and Wilhelmshaven—but the estimates of the velocity given by them range from 1·6 to 4·5 km. per sec. Neither Mallet's nor Seebach's method led to any definite result for the depth of the focus. The only estimate suggested, and that doubtfully, is one of about 11 km. or 7 miles, founded on a faulty method of Rudolf Falb (1838–1903). The conclusion which the commission regarded as the most clearly established by their labours was the coincidence of the epicentral area with the crest of a mountain, of which the southern slope, steep and faulted, is principally composed of crystalline rocks, while the gentler northern slope is formed of folded Jurassic and Neocomian rocks.

If the French commission thus added little to our knowledge of the earthquake, we are indebted to Fouqué and Lévy for some interesting experiments on the velocity of earth-waves in the surface-rocks (pp. 57–77)†. In the previous experiments of

* *Mission d'Andalousie: Études relatives au tremblement de terre du* 25 *Décembre* 1884 *et à la constitution géologique du sol ébranlé par les secousses.* Paris, *Mém. Savants Étrang.* vol. 30, 1889, no. 2, 772 pp., 42 plates.

† Summarised in Fouqué's *Les Tremblements de Terre*, pp. 219–247.

Pfaff (1825–86), Mallet, Abbot and Milne (arts. 66, 149, 198), the source of disturbance was on the surface; in the French experiments, the source was on the surface in some cases, and in mine-workings in others. In the former, the photographic records revealed a series of maxima in the disturbance that sometimes lasted for nearly two seconds; in the latter, there was only one maximum, the vibrations rapidly died away, and the record resembled that of a surface disturbance at a short distance. The principal results for the velocity are:

Granite	2·45–3·14	km. per sec.
Carboniferous sandstone ...	2·00–2·53	,,
Permian sandstone	1·19	,,
Cambrian marble	·63	,,
Sand, about	·30	,,

FERNAND DE MONTESSUS DE BALLORE

164. Since the death of Alexis Perrey in 1882, seismology in France has been almost identified with the name of Fernand Jean Baptiste Marie Bernard comte de Montessus de Ballore (1851–1923). Born on 27 Apr. 1851, Montessus received his early military training, in company with the present Marshal Foch, at the École polytechnique. In 1881, being then captain of artillery, he was sent in charge of a military mission to the republic of San Salvador, where he spent the hours of duty in drilling and instructing the native troops, and his hours of leisure in studying the earthquakes of that disturbed country. On his return to France in 1885, he became director of studies at the École polytechnique. His first important memoir—on the earthquakes and volcanic eruptions of Central America—appeared in 1888. The earthquakes of so limited a region, he soon realised, were insufficient for the deduction of general laws of distribution, and he was thus led to embark on a great undertaking, the compilation of no less than a world-catalogue of earthquakes. That Montessus was possessed of an almost "infinite capacity for taking pains" is evident, for the construction and discussion of the catalogue occupied him for more than half his working lifetime.

It would seem that, in this catalogue, individual earthquakes were entered, not chronologically, but according to countries;

for, from 1892 onwards, Montessus issued a series of 23 regional memoirs, in which he analysed the distribution of earthquakes in certain seismic areas. The whole series, revised and extended, form the substance of his first great work, that on seismological geography, published in 1906 (art. 172).

165. The following year was a turning-point in Montessus' life. In it appeared his second great work, the well-known treatise on seismology (art. 173), and, towards its close, he left France for Chili, the country in which he spent the last fifteen years of his life.

The seismological service of Chili, of which Montessus was appointed director, was one result of the great Valparaiso earthquake of 16 Aug. 1906. Montessus' early years in Chili were spent mainly in organising the new service, in establishing a central observatory of the first class at Santiago, four stations of the second, and 29 of the third, class; also in enlisting the aid of telegraph officers, lighthouse keepers, station masters, teachers, and others. From 1909, he gave lectures on construction in earthquake countries to students of architecture and engineering in the university of Santiago.

Montessus' industry showed no abatement towards the close of his busy life. In addition to his official duties, he compiled a bibliography of seismology (1915–19; art. 180), and, shortly before his death on 29 Jan. 1923, he had finished the manuscripts of two more volumes, one on seismic and volcanic ethnography, the other on seismological geology, neither of which did he live to see in print (art. 181).

Few men have given their lives to seismology so wholeheartedly and so persistently as Montessus. Some half-dozen brief papers relate to other subjects, but the vast majority of his memoirs, including his four great books, are confined to the study of earthquakes. It was the work that his hand found to do, and he did it with his might*.

166. Earthquakes of Central America. In his first scientific memoir, Montessus naturally turned to the earthquakes and

* *La Géologie Séismologique*, pp. 453–458; *Amer. Seis. Soc. Bull.* vol. 2, 1912, pp. 217–223.

volcanic eruptions of Central America*. This work was written in Spanish and was published before he left San Salvador. Though more than 2300 earthquakes and 137 volcanic eruptions are described within its pages, Montessus rightly regarded it as merely the skeleton of the new and more complete memoir† that he prepared on his return to the libraries of France. How widely he searched for his materials there is evident from the extensive bibliography given in the concluding pages (pp. 269–285).

Much of the success of this memoir was due to Montessus' analysis of Perrey's laws on the lunar frequency of earthquakes. The results of his criticism are given in an earlier section (art. 60). He also considered the annual and diurnal variations of frequency, concluding that there was no certain evidence of an annual period (pp. 19–23), but—though he afterwards abandoned this view (art. 178)—that the law of a nocturnal maximum and diurnal minimum was well-established (pp. 25–28).

The longer and more valuable part of the memoir is the catalogue of seismic and volcanic phenomena (pp. 79–267, 287–293). Neglecting the few entries dating before the Spanish conquest, there are notices of about 1300 earthquakes from 1526 to 1886—earthquakes of such importance that the time of each is recorded. Of this number, more than 90 per cent. occurred after the year 1850. Many of the shocks must have been of great strength, for San Salvador has been ruined 14 times by earthquakes, and the old town of Guatemala (1541–1773) seven times. In both memoirs, Montessus draws attention to the fact that these towns are built on the flanks of extinct, or almost extinct, volcanoes; while the new town of Guatemala and other towns close to active volcanoes have never been destroyed. The less important shocks, he notices, occurred as a rule along the recent volcanic faults that traverse the western slope of the country and are roughly parallel to the coast. Moreover, the portions of the faults that are most subject to earthquakes are far removed from active

* *Tremblores y erupciones volcánicas en Centro-America*, San Salvador, 1884. I have not been able to consult this work, but a summary of its contents is given in Paris, *Ac. Sci. C. R.* vol. 100. 1885, pp. 1312–1315.

† *Tremblements de terre et éruptions volcaniques au Centre-Amérique.* Dijon, *Soc. Scien. Natur. Mém.* 293 pp. quarto.

volcanoes and lie near the points of intersection of these longitudinal faults with secondary transversal faults (pp. 60–62).

167. Distribution of Earthquakes in Space. In one of his early papers (1896), Montessus outlines his plan to "reduce the present chaos of seismological study to some sort of order." (i) He would, in the first place, build up a world-catalogue of earthquakes more extensive than any before dreamed of. With the aid of this catalogue, he hoped (ii) to refute certain laws that were held to govern the distribution of earthquakes in time; and (iii) to define the distribution of earthquakes over the surface of the globe. On the second of these points, he never wearied of insisting (arts. 60, 175–178). The first and third are his chief contributions to seismology.

Montessus must have begun work on his great catalogue either before or soon after his arrival in San Salvador, for, in 1900, he speaks of having been engaged on it for twenty years. As it was finished in 1906, it probably occupied much of his spare time for a quarter of a century. During these years, it should be remembered, he was engaged on military duties and he wrote as many as 86 papers with a total of more than 1300 pages. Yet, during the 26 years, the catalogue never ceased to grow. By 1889, there were 45,000 entries, two years later 63,000, in 1896 more than 100,000. In 1899, the number had increased to 121,000, in 1900 to 131,000, in 1903 to nearly 160,000, and in 1906 to 171,434. The extent of the catalogue may be gathered from the fact that the MS. volumes occupy a length of 85 feet of bookshelves*.

From 1892 to 1904, while the catalogue was growing most rapidly, Montessus issued his valuable memoirs on the principal seismic regions of the world†. Europe was covered in a series of

* *Brit. Ass. Rep.* 1909, p. 61. The catalogue is now stored in the library of the Société de Géographie, Paris.

† *Ann. Mines*, vol. 2, 1892, pp. 317–328 (France and Algeria); *Arch. Sci. Phys. Nat.* vol. 28, 1892, pp. 31–39 (Switzerland); vol. 31, 1894, pp. 5–20 (Centr. Europe); vol. 33, 1895, pp. 33–61 (Italy); vol. 4, 1897, pp. 125–146, 209–230 (Japan); vol. 5, 1898, pp. 201–206 (United States); vol. 7, 1899, pp. 344–348 (Centr. Asia); vol. 9, 1900, pp. 253–268 (Mexico); vol. 11, 1901, pp. 96–98 (Oceans); vol. 13, 1902, pp. 375–395 (Erzgebirge); Batavia, *Natuurk. Tijdschr.* vol. 57, 1896, pp. 347–360 (Dutch Indies); vol. 61, 1901, pp. 40–50 (Philippines); Brux. *Soc. Belge Géol. Bull.* vol. 18, 1904, pp. 79–105 (S. Andes);

eleven memoirs (1892–1901), Asia in six (1896–1901), America in five (1892–1904), Africa is represented by Algeria (1892). Australia and New Zealand, perhaps owing to the brevity of their records, are left untouched; oceanic earthquakes are considered in a single memoir (1901). In the three memoirs published in 1892 (Switzerland, France and Algeria, and Mexico), the seismicity of the different regions into which he divided the countries, was represented by shading of various depths of tint. With the memoir on the earthquakes of Central Europe (1894), he introduced two more precise methods of depicting the distribution of earthquakes in space.

168. In the first method, he supposed that, in a well-defined seismic region of area A sq. km., p earthquakes have been recorded in n years*. On an average, then, one earthquake would be felt per year in an area of pA/n sq. km. Assuming the earthquakes to be distributed uniformly over the region considered, he drew on the map of the region two series of equidistant lines at right angles to one another dividing the map into squares, the area of each of which is proportional to pA/n, and the side to $\sqrt{(pA/n)}$. The inverse of the latter expression, he adopts as the measure of the seismicity of a region. Thus, for the British Isles, Montessus defines ten regions, the number following each region being the value of the expression $\sqrt{(pA/n)}$ in km.: (1) Scottish Lowlands, 63; (2) Perthshire and north-east coast of Scotland, 72; (3) Northern and Central England, 89; (4) English coast of the Channel, 99; (5) Caledonian Canal, 105; (6) South-eastern

Bucharest, *Inst. Meteorol. Ann.* vol. 17, 1905, pp. 57–78 (Roumania, etc.); *Geol. Soc. Quart. Jl.* vol. 52, 1896, pp. 651–668 (British Empire); *Himmel und Erde*, 1900, pp. 518–559 (Mexico); *India Geol. Surv. Mem.* vol. 35, 1900, pp. 153–194 (India); *Ital. Soc. Sism. Boll.* vol. 6, 1900, pp. 115–130 (Greece); Madrid, *Soc. Hist. Nat. An.* vol. 23, 1894, pp. 175–184 (Spain, etc.); *Mex. Soc. "Alzate" Mem.* vol. 7, 1892, pp. 49–60 (Mexico); vol. 11, 1898, pp. 263–277 (Centr. and S. America); St Pétersb. *Com. Géol. Bull.* vol. 18, 1899, pp. 195–223 (Russia); vol. 19, 1900, pp. 31–53 (Balkan Penin.); Stockh. *Geol. För. Förh.* vol. 16, 1894, pp. 225–230 (Scandinavia).

* From 1892 to 1894 that is, in the first six memoirs, Montessus took the earthquake-day as the unit (that is, all earthquakes felt in a single day were counted as one); from 1895 onwards, each shock was counted. In adopting frequency as a measure of seismicity, Montessus relied on his statement that countries in which earthquakes are especially frequent are also those in which they are most intense (*Ital. Soc. Sism. Boll.* vol. 3, 1897, pp 9–14). The inference is obviously not quite correct.

Ireland, 132; (7) Wales, 138; (8) East Anglia 174; (9) North-eastern Scotland; (10) Shetland Isles. He used this method from 1894 to 1901, but abandoned it after the latter year, having come to realise that, as earthquakes are discontinuous both in time and space, it is illogical to represent their distribution as being uniform for a given interval over a region of considerable area.

169. The second method depends on seismic centres rather than on seismic regions. When the centre of a slight earthquake is unknown, he assigned the earthquake to a centre that lies within 12½ miles of the place at which it was felt. Each centre is then represented by a circular spot of an area depending on the number of earthquakes felt within it, such number being appended to the spot. In the memoir on Japan (1897), he introduced a definite scale for the size of the spots. In successive years, the scale was slightly modified, becoming nearly stable in the year 1904*.

170. By the year 1895, Montessus had published eight of his regional memoirs; the earthquakes recorded in his catalogue fell just short of a hundred thousand in number; they were assigned to as many as 6789 centres spread over 853 more or less unstable regions. He therefore felt himself in a position to state some of the laws connecting the frequency of earthquakes with the relief of the ground†.

Expressed in their most general form, these laws are as follows:

(i) The unstable seismic regions accompany the great lines of corrugation of the earth's crust, that is, its principal features of relief, whether above or below the level of the sea.

(ii) In a group of adjacent seismic regions, the most unstable are those which present the greatest differences of relief.

* In the memoir on the South Andes (1904), there were 24 degrees in the scale, the numbers of shocks being grouped as follows: 1, 2–3, 4–5, 6–7, 8–10, 11–13, 14–17, 18–23, 24–32, 33–54, 55–69, 70–89, 90–111, 112–134, 135–159, 160–194, 195–229, 230–266, 267–306, 307–349, 350–397, 398–509, 510–864, 865–1234.

† *Arch. Sci. Phys. Nat.* vol. 34, 1895, pp. 113–133; vol. 5, 1903, pp. 640–660; Paris, *Ac. Sci. C. R.* vol. 114, 1892, pp. 933–935; vol. 120, 1895, pp. 1183–1186.

It is unnecessary to follow Montessus in his very full discussion of the particular cases of these laws. It will be seen that they do not differ greatly from those given by Mallet for the whole world in 1858 (art. 69) and by Milne for Japan in 1895 (art. 201), though they are founded on a far more extensive series of observations.

171. Some years later, in 1903, when the catalogue contained nearly 157,000 entries, Montessus was led to another general law, which he stated in the following terms*: The earth's crust trembles almost only, and nearly equally, along two narrow zones, which lie along two great circles inclined at an angle of 67°, and coinciding with the two most important lines of relief of the earth's surface. He called them the Mediterranean or Alpino-Caucasian-Himalayan-New Zealand circle, and the Pacific or Anglo-Japanese-Malayan circle, and located the positions of their poles as respectively 40° 45′ N. lat., 150° 10′ W. long., and 35° 40′ N. lat., 23° 0′ E. long. Of the total number (156,781) of earthquakes, he assigns 83,946 or 53 per cent. to the Mediterranean circle, 64,406 or 41 per cent. to the Pacific circle, and only 8429 or 6 per cent. to other regions.

Montessus was careful to point out that earthquake-regions are not distributed uniformly along the two zones. In the Pacific circle, they are clustered without interruption along an immense arc of 225° or 15,900 miles, from the Gulf of Ancud in the south of Chili to the island of Engano in south Sumatra. In the Mediterranean circle, on the other hand, they are massed along an arc of only 114° or 7900 miles, except for the isolated centre of instability in New Zealand.

In the following year, he pointed out that these two great unstable circles coincide with the geosynclinals of the secondary epoch, as outlined by Haug, transformed in the tertiary epoch into the great chains of recent elevation†.

172. Montessus adopts the remarkable law here stated as the foundation of his first great work, that on seismological geo-

* Paris, *Ac. Sci. C. R.* vol. 136, 1903, pp. 1707–1709; *Conf. Séis. Intern. 2me conf.* 1904, pp. 325–334.

† Paris, *Ac. Sci. C. R.* vol. 139, 1904, pp. 686–687.

graphy*. The regions lying outside the two great circles are, however, sub-divided into the north Atlantic areas and the extra-European continental areas, the former including Europe to the north of the Alps and the Atlantic slope of the United States and Canada, and the latter Siberia, China, India, Australia, Africa and Brazil. New Zealand is also transferred from the Mediterranean to the Pacific circle. With these changes and with the final entries in his catalogue, he obtained the following numbers:

		No. of earthquakes	per cent.
I.	North Atlantic continent	8939	5·21
II.	Extra-European continents	6343	3·71
III.	Mediterranean geosynclinal	90,126	52·57
IV.	Circum-Pacific geosynclinal	66,026	38·51

The two great synclinals thus contain 91 per cent. of the earthquakes recorded in Montessus' catalogue, and the continental areas, notwithstanding their much greater size, only 9 per cent.

Nearly every district referred to in this work had already been considered by Montessus in his various regional memoirs; but all the accounts were brought up to date (1906), and, in addition, a few regions, such as Africa, Brazil and New Zealand, were described for the first time. In his own regional maps, 56 in number, Montessus represents the seismicity by the second of the methods described above.

By this great work, Montessus made the study of earthquake-distribution his own. It was the first attempt to depict that distribution both for the whole world and for individual seismic regions, the first also to consider on a large scale its relations with the geological structure of the different countries. In the twenty years that have passed since then, no one has endeavoured to touch the subject on the same scale. If Montessus' volume is the first to be written on the seismological geography of the world, it may well be the last, for future efforts will probably aim at the discussion of the laws of distribution in detail, after the manner of that which Milne accomplished for Japan in 1895 (art. 201).

* *Les Tremblements de Terre: Géographie Séismologique*, 1906, v + 475 pp.

173. Seismological Textbooks. It is difficult to fathom Montessus' industry during the twenty-five years ending with 1907. As if the world-catalogue of earthquakes, the preparation of his regional memoirs and of his first great work, *Géographie Séismologique*, were not sufficient, he was all this time gathering that wide knowledge of earthquake-literature that culminated in 1907 in the companion volume, by far the most detailed treatise that we possess on the science of seismology*. Of the three parts into which *La Science Séismologique* is divided, the first (pp. 43–276) is devoted to macroseisms or sensible earthquakes, the second (pp. 277–406) to microseisms or instrumental earthquakes, including under this heading the registration of distant earthquakes and its bearing on the constitution of the globe, and the third (pp. 407–542) to megaseisms or destructive earthquakes and the construction of buildings in unstable countries, a subject in which Montessus never ceased to be interested. How widely he had searched for his materials is evident from the fact that he refers to as many as 340 authors.

Four years later, *La Science Séismologique* was followed by *La Sismologie Moderne*†, a popular textbook about one-quarter the size of its predecessor, and written on similar lines, except that it contains an interesting chapter (pp. 230–240) on the earthquakes of France and her colonies.

174. Frequency and Periodicity of Earthquakes. From ten years' observations (1833–1842), Perrey (1843) estimated that, on an average, 33 earthquakes occurred every year in Europe and the adjoining portions of Asia and Africa (art. 49). During the first half of the nineteenth century, Mallet (1858) recorded 3240 earthquakes, an average of 65 a year. Montessus (1895) raised the probable number considerably by an ingenious method. Recognising that the various catalogues of earthquakes differed greatly in fullness, he divided them into three classes, historical, seismological and seismographical, the earthquakes of the first class being great shocks that entered into the history of a country, and those of the second including also the slighter shocks recorded

* *La Science Séismologique*, 1907, vii + 579 pp., 31 plates.

† 1911, xx + 284 pp., 17 plates.

by interested observers and students. He divided the whole earth into 451 sub-regions, for 93 of which he possessed catalogues of two of the three classes. Denoting the annual numbers in each class by i_1, i_2, i_3, he found the following mean values:

$$i_2/i_1 = 4{\cdot}26, \quad i_3/i_1 = 26{\cdot}59, \quad i_3/i_2 = 6{\cdot}44,$$

for 44, 28 and 22 regions, respectively. These ratios were determined independently, and an interesting test of their accuracy is given by the fact that 26·59/6·44=4·18. If we assume that i_3 is the actual number of shocks, it follows that the historical documents report on an average 3·76 per cent. of the shocks, and the seismological documents 15·52 per cent. Now, the total number of shocks in each class every year are

$$\Sigma i_1 = 341{\cdot}35, \quad \Sigma i_2 = 878{\cdot}57, \quad \Sigma i_3 = 2222{\cdot}24$$

for an area of 11,691,000 sq. km. Multiplying the value of Σi_1 by the ratio 26·59 and that of Σi_2 by 6·44, the total annual number of earthquakes for the above area would be 16,957, or one every half-hour. Assuming that shocks occur at the same rate over the whole surface of the earth (510,000,000 sq. km.), this would give a mean annual number of 437,768 earthquakes, or very nearly one a minute*.

175. From the beginning of his scientific career, Montessus was impressed by the apparent constancy of seismic activity. In 1912, he extended his survey in space to the whole world and in time to the Christian era. Using Milne's catalogue of more than 4000 destructive earthquakes, he drew a graph representing the numbers in successive half-centuries. Up to A.D. 650, the curve is nearly horizontal, with an average of seven a half-century. In the next thousand years, it rises slightly, the average number each half-century being 55. From 1650 to 1850, it rises more rapidly, from about 1 to 11 earthquakes a year. But, after 1850, the curve of annual numbers becomes abruptly horizontal, and this form is maintained until 1899, when the catalogue closes, the mean annual number being about 30. For Japan, again, the curve for the last seven centuries is nearly horizontal, and

* Paris, *Ac. Sci. C. R.* vol. 120, 1895, pp. 577–579; *Beitr. Geophys.* vol. 4, 1900, pp. 345–346.

Montessus thus concludes that, for nearly two thousand years, mondial seismic activity, as measured by the number of megaseisms, has been nearly constant*.

176. The same uniformity, in Montessus' opinion, applies to shorter intervals of time. In an early memoir (1888), he considered the lunar periodicities as defined in Perrey's laws (art. 60), he returned to the subject in 1889 and 1913, and it is only recently (1922) that the validity of his reasoning has been questioned†. Later he applied his destructive criticism to the annual and diurnal periods and to various other fluctuations in earthquake-frequency.

As I have remarked before (art. 61), Montessus' criticism of Perrey's third law seems to be well-founded, though his objections to the first and second laws are less conclusive. The third law was re-considered in 1889, the number of entries in his catalogue having grown to nearly 45,000, and with the same result‡. He returned to the first law in 1913, using more than 2000 destructive earthquakes recorded in Milne's catalogue from 1792 to 1899. Dividing the lunar month into 28 intervals, he found that the maximum occurred near the interval containing full moon, and the minimum in that containing new moon—a result that gives no support to Perrey's first law, which requires a maximum both at full and at new moon§.

177. Montessus' criticisms on the annual and diurnal periodicities must next be noticed||.

For the annual period, he considers what he calls Perrey's law that earthquakes are more frequent in winter than in the other seasons, though he extends the law so as to include autumn as well as winter. Of 165 good series of earthquakes with a total of

* Paris, *Ac. Sci. C. R.* vol. 154, 1912, pp. 1843–44.

† L. A. Cotton, *Amer. Seis. Soc. Bull.* vol. 12, 1922, pp. 115–144.

‡ Paris, *Ac. Sci. C. R.* vol. 109, 1889, pp. 327–330; *Arch. Sci. Phys. Nat.* vol. 22, 1889, pp. 409–430.

§ Paris, *Ac. Sci. C. R.* vol. 156, 1913, pp. 100–102. Montessus' remarks on the above result show that he somewhat misunderstood the problem.

|| I must pass over his short papers on fluctuations in earth-frequency with variations in the latitude, in sunspot frequency, and in rainfall, in each of which he regards the fluctuations as unproven (Paris, *Ac. Sci. C. R.* vol. 147, 1908, pp. 655–656; vol. 155, 1912, pp. 379–380, 560–561; vol. 156, 1913, pp. 1194–1195).

38,967 earthquake-days, he finds that 85 series with 20,258 days follow the law, while 80 series with 18,709 days are opposed to it. Further, if M and m are the maximum and minimum number of earthquake-days per season in each of the 85 series that follow the law, and T the total number, he finds that, as T increases, the ratios M/m and $(M-m)/T$ tend to 1 and 0, respectively*. A few years later (1906), he divided his 81 series of 75,737 earthquakes into two groups, according as the regions lie to the north or south of the parallel of 45° N. lat. In the former class, the maximum falls during the winter months (October to March) in 90 per cent. of the series; in the latter class in 47 per cent. This result, he says, is easy to interpret. In the northern regions, more time is spent unoccupied and at rest during the winter months, and thus he infers that earthquakes are produced equally at all seasons†—reasoning of which we cannot but admire the boldness. Lastly, in 1913, he once more discusses the subject with the aid of Milne's catalogue of destructive earthquakes divided according to regions. The maximum occurs in eight regions in January, two in February, one in April, one in May, two in June, one in August, six in October, and two in December—that is, in 18 regions in the winter months, and five in the summer months. Further, he finds that, as T increases from 144 (Moluccas) to 3441 (whole world), the ratio $(M-m)/T$ decreases, though with some irregularity, from 9·70 to 1·42; and he suggests that it would probably tend to zero if the number of earthquakes were large enough. He therefore concludes that the frequency of megaseisms is independent of the seasons‡.

* Montessus' method of analysis is the simple one employed by Merian, Hoff and Perrey. He objected to the use of harmonic analysis for the strange reason that such a method assumes the existence of the period which it is endeavouring to prove (*La Science Séis.* 1907, p. 235).

† See *Phil. Trans.* 1893 A, pp. 1115–1116 for an estimate of the effect of this condition.

‡ Paris, *Ac. Sci. C.R.* vol. 112, 1891, pp. 500–502; vol. 143, 1906, pp. 146–147; vol. 156, 1913, pp. 414–415; *Arch. Sci. Phys. Nat.* vol. 25, 1891, pp. 501–517. I ought perhaps to give in a footnote my reasons for dissenting from Montessus' conclusions. (i) In Japan and several other countries, the annual periodicity is more marked for weak, than for strong, earthquakes (*Phil. Trans.* 1893 A, pp. 1116–1120). (ii) The law which Montessus attributes to Perrey is not a correct statement of the annual periodicity, the maximum epoch of which occurs, as Montessus remarks, in various months, though most often in winter. (iii) Large areas, such as America, Asia, the

178. The daily periodicity, or, rather, the supposed nocturnal predominance of earthquakes, is treated in a similar manner. Denoting by d the number of earthquakes recorded during the twelve hours of the day and by n the number of those during the twelve hours of the night, Montessus obtains the values of the ratio d/n for different kinds of catalogues. In regions poorly defined from a physical point of view, like France, the mean value of the ratio is 0·75. For local series of short duration it is 0·79, and, for those of long duration, 0·76. Series of seismological commissions give a mean of 0·82. Montessus remarks that the ratio increases (from 0·75 to 0·82) with the increasing scientific value of the group. This he considers a strong presumption in favour of a value really equal to unity. Further, the more favourable conditions prevailing during the night hours would concern only shocks of weak intensity. For degrees 10 and 9 of the Rossi-Forel scale, the ratio d/n is slightly greater than unity. It decreases regularly to 0·65 for the degree 5, 0·67 for the degree 4, and 0·60 for the degree 3, after which it rises to 0·73 for the degree 2 and 1·80 for the degree 1. The last group of observations—those recorded instrumentally at the Italian stations—form an exception to the tendency of the ratio towards unity, its value being 2·04 at Rome and 1·73 at Velletri, and its mean value 1·49. Montessus, however, attributes this preponderance during the day to artificial disturbances, for the hourly curve of the group rises regularly from 7–8 a.m. to 10–11 a.m. He thus comes to the conclusion—and it is again a bold one—that earthquakes occur with equal frequency throughout the day*.

Northern Hemisphere, and the World, contain some regions with a winter, and others with a summer, maximum. The effect of including both is of course to reduce the amplitude of the annual period for the sum of the regions. Take, for instance, the records of 16 important stations in different parts of Japan. In the north-east of the country, the maximum epoch falls in May, June or July; in other parts, most frequently in December or January. Taking the mean monthly numbers of earthquakes at each station as unity, the amplitude of the annual period is as high as 0·76 in the former region and 0·75 in the latter, the average values being respectively 0·36 and 0·30. But, if we group together the earthquakes recorded at all the 16 stations (reducing them to their equivalent numbers for equal durations of records), the amplitude is only 0·11, while the maximum epoch falls in May.

* Paris, *Ac. Sci. C. R.* vol. 109, 1889, pp. 327–330; *Arch. Sci. Phys. Nat.* vol. 22, 1889, pp. 409–430. It should be noted that, for ordinary earthquakes

179. Miscellaneous Phenomena. Somewhat late in life (1912), Montessus' doubts spread to the existence of isoseismal lines and the value of intensity scales. The tests in some of these scales are certainly open to criticism. But, even assuming their exactness, Montessus claims that the estimates of the intensity in any one place are so variable that "it is impossible and misleading to trace isoseismal curves." Such curves, he thought, have the grave defect of leading to false conceptions, as for instance, that of earthquakes with multiple epicentres*. The only isoseismal lines that he would have traced are the boundaries of the disturbed area, of the region of greatest damage, and of that of minor damage†.

Montessus seems to stand on firmer ground when he doubts the existence of luminous phenomena accompanying earthquakes. He examines the 148 cases collected by Galli‡, and numerous observations in connexion with two recent earthquakes—the Valparaiso earthquake of 16 Aug. 1906 and the South German earthquake of 16 Nov. 1911. The descriptions he regards as generally lacking in precision; the time-intervals between the lights and the earthquakes are variable, in some cases amounting to hours; the luminous phenomena are reported from very distant stations as well as from places in the epicentral area, one of those in the Valparaiso earthquake being more than 450 miles from the centre of the meizoseismal area; they are connected more frequently with the atmosphere than with the ground. During the night of the Valparaiso earthquake, a great storm raged in Chili; on the night of 16 Nov. there is a shower of meteors with the

recorded instrumentally, the maximum of the diurnal period occurs about noon. This would account for the morning rise in the hourly curve of the Italian group.

* Montessus refers especially to his failure to draw satisfactory isoseismal lines for the Valparaiso earthquake of 1906. This was due partly to inexperience, chiefly perhaps to the small number of observations at his disposal and to the fact that the estimates of intensity were made by the observers themselves.

† Paris, *Ac. Sci. C. R.* vol. 154, 1912, pp. 1461–1463; *Amer. Seis. Soc. Bull.* vol. 6, 1916, pp. 227–231. It is unnecessary to follow Montessus in his criticism of what he calls isosphygmic lines or curves of equal earthquake-frequency. He regards them as both useless and non-existent owing to the discontinuity of earthquake-phenomena in time and space (Paris, *Ac. Sci. C. R.* vol. 133, 1901, pp. 455–457; *Beitr. Geophys.* vol. 5, 1902, pp. 467–485).

‡ *Ital. Soc. Sism. Boll.* vol. 14, 1910, pp. 221–448.

radiant-point in Ursa Major. Montessus thus concludes that if, at present, we cannot affirm or deny the existence of luminous earthquake-phenomena, all the facts known to us point to a negative conclusion*.

180. Later Works. During his residence in Chili, Montessus wrote two important works—a bibliography of memoirs relating to earthquakes and a treatise on seismological geology.

(i) Few, if any, men have been so widely acquainted as Montessus with the literature of seismology and none so competent as he to compile a bibliography of the subject. The value of his work will be evident from the statement that it contains the titles of 9140 books and papers by 1500 authors. It was issued in seven parts during the years 1915–17†. The first part contains the titles of memoirs on seismological theories, the geological effects of earthquakes, etc.; the second on northern and central Europe; the third on the circum-Mediterranean countries; the fourth on Asia, Africa and Oceania; the fifth on America, the Oceans, etc.; the sixth on accessory phenomena, relations with other phenomena, seismic architecture, etc.; while the seventh part contains supplements and additional matter.

In 1916, that is, before the bibliography was quite complete, Montessus made an estimate of the relative contributions of different nations to the advance of seismology. Out of a total of 8500 memoirs, he found that workers in Italy had contributed 2002, in France 1768, in Germany 1185, in Great Britain 911, in the United States and Canada 636, and in Japan 352, or, respectively, 23·6, 20·5, 13·9, 10·7, 7·5 and 4·1 per cent. of the whole‡.

181. (ii) The writing of *La Géologie Sismologique* occupied the last months of Montessus' life. He lived to see the text finished, but the correction of the proofs and the choice of the illustrations fell to other hands§. Seismic geology, according to Montessus,

* Paris, *Ac. Sci. C. R.* vol. 154, 1912, pp. 789–791; vol. 158, 1914, pp. 749–751; *Ital. Soc. Sism. Boll.* vol. 16, 1912, pp. 77–102; *Amer. Seis. Soc. Bull.* vol. 3, 1913, pp. 187–190.

† *Bibliografia general de Tremblores y Terremotos*, Santiago.

‡ *Ital. Soc. Sism. Boll.* vol. 20, pp. 263–272.

§ 1924, xiv + 488 pp. The book was edited by his brother R. de Montessus de Ballore, and the biographical notice and the list of memoirs were written by Armand Renier, the director of the Geological Service of Belgium.

is for the most part a discussion of the great earthquakes, more numerous than is generally supposed, during which permanent changes in the solid crust have occurred.

In 1878, R. Hoernes divided earthquakes into three classes—tectonic, volcanic and rock-fall earthquakes (arts. 119, 134). Montessus preferred two main classes—glyptogenic or geological and external dynamic earthquakes. The former were further re-arranged into epeirogenic, tectonic, and epeirogenic and tectonic, earthquakes, according as the surface displacements connected with them are vertical, horizontal, or vertical and horizontal; and the latter into volcanic earthquakes and rock-fall earthquakes (tremblements de terre d'écroulement). Far outnumbering either class are the slight earthquakes which produce no apparent individual effects on geological structure. These he left unconsidered, for he regarded their connexion with glyptogenic earthquakes as nothing more than a plausible induction.

Among the examples of epeirogenic earthquakes, Montessus took the New Madrid earthquake of 1811, the Assam earthquake of 1897 and the Kangra earthquake of 1905. To these, he would have added the Chilian earthquakes of 1822, 1835 and 1837, had he felt sure that they were accompanied by real elevations of the land; but, on this point, he accepted Suess' arguments (art. 133). The Owen's Valley (California) earthquake of 1872, the Sumatra earthquake of 1892, and the Californian earthquake of 1906 are regarded as tectonic earthquakes; and the New Zealand earthquakes of 1848 and 1855, the Mino-Owari earthquake of 1891, and the Alaskan earthquakes of 1899 as epeirogenic and tectonic earthquakes. The Hawaiian earthquake of 1868 and the Ischian earthquake of 1883 are taken as typical volcanic earthquakes; while the Port Royal earthquake of 1692, the Kingston earthquake of 1907, and the Messina earthquake of 1908 are connected with the subsidence of coasts and submarine talus.

The detailed descriptions occupy five-eighths of the book. In the remaining part we have chapters on such subjects as the isostatic readjustment of alluvial plains, avalanches, and earthflows, the migration of epicentres, and the distribution of earthquakes over the surface of the globe.

Compared with Montessus' other great works, the third and

last shows no diminution of interest, no sign of failing energy. To have produced it at the age of 71 is surely a worthy ending to a life of unresting labour in the cause of science.

182. Conclusion. From near the outset of his career (1888), Montessus' work fell into two main divisions, that were maintained for a quarter of a century. In the first, which includes his chief contributions to the science, were his numerous memoirs on the geographical distribution of earthquakes; in the second, those in which he criticises the work of other seismologists. Most of that criticism seems to me ineffective, for reasons that I have already given, and one can only regret now that time was so spent that might have been devoted to other and more useful ends. In the last ten years of his life, he added a third division, consisting of his vast, but too little-known, bibliography of seismology, and of several historical essays.

Putting on one side his critical papers, Montessus, in his active life, has given us much that we could ill afford to lose. Before his three great works, there are yet, it is to be hoped, many years of usefulness. Though unpublished, his catalogue of 171,434 earthquakes remains and is accessible to students. The results that he obtained from the discussion of this catalogue, his monographs on the different seismic regions of the world, the laws that he stated with regard to the distribution of earthquakes in space, surely form a sufficient and lasting monument of what may be accomplished by the persistence and patience of one man.

CHAPTER X

JOHN MILNE

183. During the period of about twenty years that included that of Mallet's death (1881)—say, ten years before and ten after—the study of earthquakes made rapid progress. Among the more prominent contributors were De Rossi and Mercalli in Italy, Seebach in Germany, Suess and Hoernes in Austria, Montessus in France and Dutton in the United States. Their work, however, was mainly carried out on the old lines. For the introduction of new methods of study and of a new spirit infused into seismology, we are indebted to the small band of early British teachers in Japan, to J. A. Ewing (1855–), T. Gray (1850–1908), and, above all, to J. Milne (1850–1913). In the new epoch, then opening, when seismology demanded the whole energy of its supporters as well as their active cooperation, it is not, I think, too much to claim that Milne lifted the science to an altogether different and higher plane.

184. Born at Liverpool on 30 Dec. 1850, John Milne was trained as a mining engineer under Warington Smyth (1817–90) at the Royal School of Mines. After gaining experience in the mines of Cornwall, Lancashire and Central Europe, he spent two years in an inquiry into the mineral resources of Newfoundland, while his interest in geology was shown by the valuable remains of the great auk which he brought home from Funk Island. In 1874, he acted as geologist in Beke's expedition sent out by the Royal Geographical Society to fix the exact site of Mount Sinai. A year later, he received the appointment that determined the bent of his future life, that of professor of geology and mining in the Imperial College of Engineering at Tokyo.

It was in keeping with Milne's energy and wide interests that he preferred to approach his new home by a solitary and toilsome journey overland. Crossing Asia, almost along the line of the present Siberian railway, and making many geological observations on the way, he reached Tokyo early in 1876 after the lapse of eleven months, his first night in the city giving him an introduction to Japanese earthquakes.

For the next four years, however, the direction of Milne's study was undecided. He was fully occupied, perhaps, with the duties of his new post and with the task of working up the materials collected during his travels. Many of his early papers consist of geological notes on the countries that he had visited—Newfoundland, Egypt, North-west Arabia, and Siberia. Then come a series of notices inspired by his experiences in Japan, on stone implements and other prehistoric remains in that country, and, veering gradually towards his future work, on the form and on the geographical distribution of volcanoes. Indeed, almost the only phenomena that could in after years divert Milne from earthquakes were those provided by the volcanoes of Central Japan.

185. Precursors of the Seismological Society of Japan. Soon after Milne's arrival in Japan, the earthquakes of the country received their first scientific study. In the year 1878, three interesting papers were published. The destructive earthquakes were considered by I. Hattori (1851–)*. Between the years 416 and 1872, there were 149 such earthquakes, the numbers in successive centuries being 1, 1, 7, 7, 28, 11, 10, 1, 7, 8, 15, 8, 15, 13 and 16, an average, as he remarked, of one severe earthquake in the country every ten years. Hattori noticed that severe earthquakes show a tendency to occur in clusters. There were, for instance, 39 destructive earthquakes in eight periods amounting to 46 years, or roughly, six such earthquakes in seven years. In a useful table, Hattori gave for the first 14 months the monthly numbers of after-shocks of the great earthquake of Ansei (1854), the total number being 819, of which 12 were of the first magnitude, 111 of the second and 696 of the third.

In the second paper, Edmund Naumann drew up a list of Japanese earthquakes and volcanic eruptions between the same limiting dates. The earthquakes are 405 in number, and the list of course includes many besides the destructive shocks†.

* Japan, *Asiat. Soc. Trans.* vol. 6, 1878, pp. 249–275.
† *Deutsch. Ges. Ostasien Mitth.* vol. 2, 1876–80, pp. 163–216.

W. S. Chaplin discussed the lunar periodicities of 143 earthquakes recorded at the Meteorological Observatory of Tokyo from July 1875 to Jan. 1878. While admitting that the number is too small to give definite results, he concluded that they lend no support to the laws enunciated by Perrey, for there are no more earthquakes at new and full moon, when the moon is in perigee or crossing the meridian than at other times*.

186. In 1873, the Japanese Government founded the Imperial College of Engineering at Tokyo, which for a time was the largest technical college in the world. When Milne joined the staff, he found there as colleagues William Edward Ayrton (1847–1908) and John Perry (1850–1920), the former living in Japan from 1873 to 1878 and the latter from 1875 to 1879. Soon after Perry's arrival, the two formed a scientific partnership that lasted until 1891, Ayrton being "the worldly practical member of the firm, Perry the dreamer," Ayrton "a man of restless energy and of the most varied capacities," Perry "a warm-hearted Protestant Irishman, impulsive and enthusiastic."† While in Japan, they wrote two papers dealing with earthquakes, one on a neglected principle that may be employed in earthquake measurements, the other on structures in an earthquake country‡. After their return to England, their active interest in earthquakes ceased, though Perry wrote two brief notes, probably at Milne's request§.

Thus the ground was to some extent prepared when, on 22 Feb. 1880, a semi-destructive earthquake occurred near Yokohama, the strongest felt in Tokyo since the arrival of the British teachers. The earthquake possesses an interest far beyond its strength, for it led to the foundation of the first society devoted exclusively to the study of earthquakes and volcanoes.

187. The Seismological Society of Japan. On Milne's initiative, a public meeting was called at which the society was

* Japan, *Asiat. Soc. Trans.* vol. 6, 1878, pp. 353–355 with tables.

† *Roy. Soc. Proc.* ser. A, vol. 85, 1911, pp. i–viii; *Nature*, vol. 105, 1920, pp. 751–753.

‡ Japan, *Asiat. Soc. Trans.* vol. 5, 1877, pp. 181–202; *Phil. Mag.* vol. 8, 1879, pp. 30–50, 209–217.

§ *Japan Seis. Soc. Trans.* vol. 3, 1881, pp. 103–106; *Brit. Ass. Rep.* 1896, pp. 218–219.

started in the spring of 1880, with I. Hattori as president, Milne as vice-president, and W. S. Chaplin as secretary. Only three lists of members have been published, their numbers up to 1883 lying between 110 and 120. In 1882, General G. Yamada, minister of the interior, became president, J. A. Ewing, and afterwards G. Wagener, vice-president, Milne foreign secretary and D. Kikuchi (1855–1917) Japanese secretary. But, though small in numbers, the lists of members contained some names that afterwards became well known in earthquake circles. Besides those already mentioned, there were, among foreigners, T. Gray (1850–1908), E. Knipping, T. C. Mendenhall (1841–1924) and C. D. West, and later C. G. Knott (1856–1922); among Japanese, S. Sekiya (1855–96) and F. Omori (1868–1923).

Of the above active members, Sekiya and Omori form the subjects of the next chapter. Of some of the others brief notices may now be given.

188. James Alfred (now Sir Alfred) Ewing (1855–) devoted much time to the construction and use of seismographs. He was educated at Edinburgh university. From 1878 to 1883, he was professor of mechanical engineering and physics in the Imperial University, Tokyo, and, on his return to this country, he held similar posts in University College, Dundee, and Cambridge university. Since 1916, he has been principal and vice-chancellor of Edinburgh university. In addition to the invention of seismographs, Ewing wrote an exhaustive monograph on *Earthquake Measurement* published in 1883*.

189. Thomas Gray (1850–1908) was another of the pioneers in the construction of seismographs. He was educated at Glasgow university, where he studied under Kelvin (1824–1907). The years 1879–81 were spent in Japan as professor of telegraphic engineering in the Imperial University, Tokyo. After some further years at Glasgow, he was appointed in 1888 professor of dynamic engineering in the Rose Polytechnic Institute, Terre Haute, Indiana, a post which he held until his death twenty years later. If his work in England and America was chiefly devoted to physical problems, his contribution to seismology

* Tokyo, *Univ. Sci. Dept. Mem.* 92 pp., 23 plates.

during his brief residence in Japan can hardly be too highly valued*.

190. When Ewing left Japan in 1883, he was succeeded by Cargill Gilston Knott (1856–1922). Like Ewing, Knott studied at Edinburgh university. From 1879 to 1883 he was Tait's assistant there, then from 1883 to 1891 professor of physics in the Imperial University, Tokyo. While in Japan, Knott contributed three papers on earthquakes to the Seismological Society†, took part in a magnetic survey of the country, and studied the eruption phenomena of Bandaisan in 1889. On his return home, he became lecturer, and afterwards, reader, in applied mathematics in Edinburgh university, and in 1912 general secretary to the Royal Society of Edinburgh. Of a "kindly, painstaking, cheerful character," it was congenial work to him to edit the collected papers of P. G. Tait and of John Aitken, to the former of which he contributed a sympathetic biography of his old teacher. To the last, he retained his interest in seismology. In 1905–06, he gave a course of lectures at Aberdeen on the physics of earthquake phenomena. They were published in 1908, and give an admirably clear account of the physical aspects of the science‡. To the Royal Society of Edinburgh, he also contributed three important papers on seismic radiations, in the last of which he traced the paths of the primary and secondary waves through the body of the earth, and concluded that, at a depth of between one-half and three-fifths of the earth's radius, the elastic solid shell gives place to a non-rigid nucleus§.

191. In later years, Milne often claimed that the foundation of the Seismological Society of Japan marks an epoch in the

* The above details are taken from a memorial notice published by the Rose Institute, for a copy of which I am indebted to the courtesy of the Registrar, Miss Mary Gilbert.

† *Japan Seis. Soc. Trans.* vol. 9, 1886, pp. 1–20; vol. 12, 1888, pp. 115–136 (afterwards reprinted and enlarged in *Phil. Mag.* vol. 48, 1899, pp. 64–97, 567–569); and vol. 15, 1890, pp. 41–45.

‡ *The Physics of Earthquake Phenomena*, Oxford, 1908, 283 pp.

§ *Edin. Roy. Soc. Proc.* vol. 28, 1908, pp. 217–230; vol. 30, 1909, pp. 23–37; vol. 39, 1920, pp. 157–208. For the notice of Knott's life, see *Roy. Soc. Proc.* ser. A, vol. 102, 1923, pp. xxvii–xxviii.

history of seismology, and none will dispute the justice of the claim. Little, it was recognised, could be done without the aid of an accurate seismograph, and it was to the early members of the society that we are indebted for the first instruments deserving the name*. In the short time that remained before they left Japan, Gray contributed accounts of his seismographs for vertical motion and large motions and studied the theory of steady-points for earthquake-measurements†; while Ewing described his horizontal and vertical motion seismographs and his duplex pendulum seismometer‡.

For twelve years the Seismological Society carried on its useful work, coming to an end in 1892. Sixteen volumes of *Transactions* were published. They were succeeded by four volumes of the *Seismological Journal of Japan*, edited by Milne, the twenty volumes altogether containing more than 3000 pages. Counting abstracts of official papers and translations, Milne wrote almost exactly two-thirds. The remaining third consisted of 68 papers written by 37 authors §. But Milne's labours did not end with his actual contributions. It was under his guidance and led by his enthusiasm that many of the other papers were written, and that investigators were trained to carry on the work after his return to Europe.

192. Milne's study of earthquakes extended over an interval of about 33 years. (i) From 1880 to 1892, or during the existence of the Seismological Society, he was mainly concerned with ordinary earthquakes as they are felt within their disturbed areas. (ii) The next three years (1892–95), while he was editing the *Seismological Journal*, formed a period of transition; his attention was divided between local and distant earthquakes;

* The history of these seismographs is discussed in a series of letters in *Nature*, vol. 35, 1887, pp. 36, 75–76, 126, 172–173, 197, 198, 559–560, 606.

† *Japan Seis. Soc. Trans.* vol. 3, 1881, pp. 1–8, 137–139, 143–144.

‡ *Japan Seis. Soc. Trans.* vol. 2, 1880, pp. 45–49; vol. 3, 1881, pp. 140–142; vol. 5, 1883, pp. 89–91. The first and third of these instruments are also described in *Roy. Soc. Proc.* vol. 31, 1881, pp. 440–446; vol. 44, 1888, pp. 395–396.

§ The most important of these authors were Ewing, Gray, Knott, Omori and Sekiya, with respectively, 10, 6, 3, 5 and 6 papers, Omori and Sekiya also contributing a joint paper.

he was experimenting with the horizontal pendulum, afterwards known as the Milne seismograph. (iii) From 1895 until his death in 1913, or during his second residence in England, his time was chiefly devoted to the observation of distant earthquakes, with occasional digressions on the distribution of great earthquakes in space and time.

In 1895, Milne left Japan, the close of his residence there being unfortunately marked by the burning of his library and seismological observatory. He reached England in July, and, with his Japanese wife, made his home at Shide Hill House, near Newport, in the Isle of Wight.

From 1880 onwards, Milne's work was aided by grants from the British Association. The reports which he wrote as secretary of the committee on the Earthquake Phenomena of Japan contain useful summaries of the work done by himself and others in that country. At the first meeting of the Association (1895) held after his arrival in England, this committee and the Earth Tremors Committee were merged into the Seismological Committee, Milne continuing to act as secretary until his death, after a short illness, on 31 July 1913.

Though Milne contributed most of his work to scientific journals, he also wrote two well-known volumes on *Earthquakes* (1886) and *Seismology* (1898) for the International Scientific Series. In addition to nearly 2000 pages in the Japanese *Transactions* and *Journal*, he wrote the fifteen reports of the committee on the Earthquake Phenomena of Japan, and the greater part of eighteen reports of the Seismological Committee, besides many other papers in various journals.

Milne differed from Perrey and Mallet in his preference for working with others. He was not a solitary student. Retaining to the last the cheerfulness of youth, gifted with an abiding sense of humour, kindly and hospitable, possessing not only enthusiasm but also the power of kindling in others the same eagerness, abounding in energy and loving hard work, Milne was admirably adapted—no man better—to act as a founder in the now widening science of seismology*.

* *Geol. Mag.* vol. 9, 1912, pp. 337–346; *Roy. Soc. Proc.* vol. 89, 1914, pp. xxii–xxv; *Amer. Seis. Soc. Bull.* vol. 2, 1912, pp. 2–7.

193. The chief difficulty in summarising Milne's work is its wide range. Michell wrote but one memoir. Mallet's work, though it covered about sixteen years of an active life, was described in less than twenty papers and one book. Milne penetrated into nearly every corner of seismology, and his papers, if collected, would fill many volumes. I will endeavour to group his work under various headings, but, to do this within a limited space, minor details must be omitted, and a strictly chronological order to some extent abandoned.

194. Investigation of Earthquakes. Milne was a student of earthquake phenomena, rather than of earthquakes, and the Yokohama earthquake of 22 Feb. 1880 is the only one that he can be said to have investigated. At this early date, the disadvantages under which he worked were great. The Japanese public were not then educated in the observation of earthquakes. De Rossi's first scale of seismic intensity was devised in 1874, but it was probably unknown outside Italy. Milne had therefore to depend on observations of the time and direction of the shock, the latter alone being found to possess any value. These enabled him to fix the position and form of the epicentral area. To determine the depth of the focus, he measured seven angles of emergence at Tokyo and Yokohama from the horizontal and vertical motions given by various instruments, and concluded that the most probable values of the resulting depths are from 1½ to 5 miles (art. 44). Two other subjects on which Milne collected many observations were also considered—the rotation of chimneys and gravestones and the nature of the damage to buildings. In the Yokohama cemetery, for instance, the gravestones are arranged in parallel rows, and the stones were rotated in one direction only. This uniformity of rotation is clearly inconsistent with the explanation given by Mallet (art. 65), and Milne preferred one suggested to him by his colleague T. Gray. The other subject, then and afterwards, was constantly before Milne's mind, and in this paper we see his first attempt to provide rules for the construction of earthquake-proof buildings*.

* *Japan Seis. Soc. Trans.* vol. 1, pt. 2, 1880, pp. 1–116.

195. In this section may also be included the paper on the Peruvian earthquake of 9 May 1877 (art. 127), though it refers mostly to the sea-waves that were propagated across the Pacific to the coasts of Japan. The position of the epicentre was ascertained from observations on the time of occurrence, and one of the chief conclusions of the paper is that the mean depths, calculated from the measured velocities by means of the formula $h=v^2/g$, are less than those obtained from soundings. Milne suggests that the latter figures may be excessive, but it has been shown that the inequality must result from variations in the depth of the ocean*.

196. Construction of Seismographs. Though the chief share in the design of accurately recording seismographs fell to Ewing and Gray, we are indebted to Milne for the testing of the various instruments and for many improvements in detail. Probably no other seismologist has had so wide an experience as he of the seismograph in all its different forms. He wisely began with the simple apparatus in use at Comrie forty years before (art. 44), and with those recommended, but not tested, by Mallet in the *Admiralty Manual* (art. 71); but common pendulums, vessels filled with liquids, bodies floating in water, columns cylindrical and otherwise, and vertical springs were all, for one reason or another, found wanting. For several years (1875–85), the Tokyo earthquake record was made by a Palmieri seismograph (art. 87). In the latter year, this instrument was replaced by a Gray-Milne seismograph, a combination of horizontal motion pendulums of the Ewing type and a vertical motion seismograph, all recording in ink on a continuously moving strip of paper, so that the register could be continued without interruption except for the brief interval during which the paper was being changed†.

In 1880, Gray had re-discovered the type of horizontal pendulum first devised by Gerard in 1851‡, the heavy mass being

* *Japan Seis. Soc. Trans.* vol. 2, 1880, pp. 50–96; *Phil. Mag.* vol. 43, 1897, pp. 33–36; vol. 50, 1900, pp. 579–584. A similar inequality had been noticed by W. J. L. Wharton in the case of the Krakatoa sea-waves of 26–30 Aug. 1883 (*Eruption of Krakatoa and Subsequent Phenomena*, 1888, p. 148).

† *Japan Seis. Soc. Trans.* vol. 12, 1888, pp. 33–46.

‡ In principle, the horizontal pendulum had been discovered many times between 1832 and 1880—by L. Hengeller or Hengler in 1832, A. Gerard in

carried on a horizontal rod, the pointed end of which rests in a conical hole in an upright frame, while the mass is supported by a wire attached to a point of the frame nearly but not quite vertically above the pointed end of the rod. Realising the value of such a pendulum for the registration of distant earthquakes, Milne, before he left Japan (1892), invented the instrument afterwards widely known as the Milne seismograph*. With various modifications, it became for many years the standard instrument of the Seismological Committee of the British Association, and with it a vast amount of useful work was done. It is now being replaced by the much improved Milne-Shaw seismograph†.

197. Nature of Earthquake-Motion. One can well understand the interest of the early meetings of the Seismological Society when the first records were exhibited by Ewing and Milne. Most of the shocks in those days were comparatively slight, but they were strong enough to reveal the general character of the motion. Some of the results were unexpected, especially the small amplitude of the movement, which, in ordinary earthquakes, is often less than a millimetre and seldom exceeds a few millimetres, while the maximum acceleration—calculated on the supposition that the motion is harmonic—usually ranges from 10 to 100 mm. per sec. per sec. Sometimes, it reaches 500 mm. per sec. per sec. and the shock is then on the verge of being dangerous. The main conclusions are now well known and a brief summary will be sufficient. The earthquake begins with a series of preliminary tremors, in which the amplitude is a small fraction of a millimetre and the period from $\frac{1}{25}$ to $\frac{1}{16}$ of a second. They may be followed by a shock consisting of three or four back-and-forth motions, with an amplitude of 1 to 10 mm. and a period usually less than one second. The end portion is a series of irregular motions, in which the greatest amplitude ranges from a fraction of a millimetre to one or two millimetres, and the period from

1851, Perrot in 1862, M. H. Close and F. Zöllner in 1869, C. Delaunay in 1871, Lord Kelvin in 1879, and J. A. Ewing and T. Gray in 1880 (*Brit. Ass. Rep*. 1895, pp. 184–185).

* *Brit. Ass. Rep*. 1896, pp. 187–188; 1897, pp. 137–145; 1904, pp. 43–44.

† *Brit. Ass. Rep*. 1915, pp. 57–58.

$\frac{1}{6}$ to $\frac{1}{3}$ of a second. As the movement dies away, the period increases, and waves with a period of two or three seconds have been recorded. An interesting point is that the direction of these irregular vibrations is constantly changing; it seems to have no connexion with the direction of propagation, whereas the direction of the principal vibrations coincides approximately with that direction*.

198. Seismic Experiments. The registration of earthquakes, it might at first sight be expected, would give all the information desired on the nature of earthquake-motion. But the earthquakes came without warning and from different regions. The observing stations were limited in number. Above all, that source was usually of considerable size, and distortional vibrations from the nearer parts of the focus coalesced with condensational vibrations from more distant portions. It was necessary, therefore, to simplify the disturbance, and this was attained by explosions of dynamite and by dropping a heavy weight (1710 lb.) from various heights up to 35 feet.

The idea of making such experiments occurred to Milne in 1880†. Altogether ten series were made from 1881 to 1883, three of them in 1881 in conjunction with T. Gray.

The most interesting result of the experiments was the complete separation of the condensational and distortional vibrations. When the record was made with a single index on a fixed glass plate, the index first moved in a straight line away from the origin. After this, it was suddenly deflected and yielded a diagram in which the movement was elliptical or in the form of the figure 8, the two sets of vibrations being then compounded. When, however, two bracket seismographs of the Ewing type were installed, one to record the movement in the direction of the source, and the other in the perpendicular direction, the vibrations of the two types were kept apart, and the variations of each could be studied. Near the origin, the amplitude of the condensational vibrations was the greater, but died out more rapidly than that of the others with increasing distance from the origin. In both

* *Brit. Ass. Rep.* 1881, p. 202; 1882, pp. 208–209; 1884, pp. 243–244.

† The earliest suggestion of such experiments is probably that by Honoratus Faber in 1670 by exploding gunpowder buried in pits (art. 7).

types of vibration, the period increased as the waves diverged. In the velocity of the waves, Milne found many variations. It increased, for instance, with the strength of the disturbance, and, in any one artificial earthquake, it diminished as the waves spread outwards. In the condensational vibrations, the mean velocity was 438 ft. per sec.; in the distortional vibrations, 357 ft. per sec. The effect of inequalities in the ground was also studied. Small hills seemed to interpose but little obstacle to the waves; excavations, such as ponds, checked the passage of both types of vibration*.

A matter of some practical consequence brought out by these experiments is that two points of ground only a few feet apart did not always synchronise in their movements. This result led Milne to make some further observations during earthquakes, and he was able to show that points only three or four feet apart experienced slightly different motions. He also made a survey of about nine acres in the grounds of the Imperial College of Engineering at Tokyo. On the west side, the ground is somewhat marshy; elsewhere it is dry and hard. He found that, as is now well known, the maximum acceleration (which measures the intensity of the shock) was greater on the soft than on the hard ground. An unexpected result was that the acceleration at the bottom of a pit ten feet deep was much less than at the surface†.

199. Catalogues of Japanese Earthquakes. During his residence abroad, Milne compiled two catalogues of Japanese earthquakes, one showing the distribution of seismic activity in past time, the other the distribution throughout the country within a limited and recent period. The earlier catalogue is less well known than it deserves to be. It contains a list of 366 great earthquakes, from 295 B.C. to A.D. 1872, and gives for each shock the date, the intensity (according to a rough scale of two degrees), and the district chiefly affected. One result obtained from this catalogue is worth noticing. The increase in the number of entries towards the present day, which is so characteristic a feature of all

* J. Milne and T. Gray, *Phil. Trans.* for 1882, pp. 863–883; J. Milne, *Japan Seis. Soc. Trans.* vol. 8, 1885, pp. 1–82.

† *Japan Seis. Soc. Trans.* vol. 10, 1887, pp. 1–36; vol. 12, 1888, pp. 63–66, 67–75; vol. 13, 1890, pp. 41–89.

other catalogues extending over many centuries, is not observable here from the seventh century onwards. The probable explanation, according to Milne, is that Japan has remained throughout in the same state of civilisation, while other countries have emerged from a state of semi-barbarism*.

Soon after the foundation of the Seismological Society, Milne began the systematic study of the distribution of Japanese earthquakes. He adopted the plan of distributing bundles of postcards among the Government offices in all important towns within a hundred miles of Tokyo. Every week a postcard was to be returned to him with notes of any earthquakes observed. The results of two years (1881–83) were enough to show the efficiency of the method—387 earthquakes were recorded during this time in north Japan—and to reveal the principal laws of distribution. The "barricade of postcards" was then extended northwards as far as Sapporo, 450 miles from Tokyo, but the labour and expense were becoming too great to be borne by one man, and the Imperial Meteorological Department was induced to carry on the work. The materials for the eight years 1885–92 were afterwards placed in Milne's hands, and thus enabled him to prepare his great catalogue of 8331 earthquakes recorded in Japan during these years†.

It is difficult to appreciate the labour involved in the preparation of this catalogue. For every earthquake, the number of postcards received varied from three or four to several hundred, and it is estimated that the total number of documents examined must lie between eighty and a hundred thousand. For each earthquake, a separate map was drawn, this being the work of the Meteorological Department. The reduction of the great mass of material within convenient limits, the determination of areas and epicentres were the work of Milne and an assistant. For convenience in printing, the catalogue is in two parts. In the first, are given the time of occurrence of every earthquake, the extent of land-area shaken, and figures which, by reference to

* *Japan Seis. Soc. Trans.* vol. 3, 1881, pp. 65–102. Since this catalogue was published, a more extensive "Earthquake Investigation Committee Catalogue of Japanese Earthquakes" has been compiled under the superintendence of the late S. Sekiya (art. 220).

† *Seism. Jl. Japan*, vol. 4, 1895, pp. i–xxi, 1–367.

the index-map, determine the approximate position of the epicentre and the boundary of the disturbed area. The second gives the lengths of the axes of that area, the distance of the epicentre from the shore when the earthquake was submarine, and the seismic district to which the earthquake belongs.

200. With these catalogues may perhaps be included the valuable list of earthquakes recorded at Tokyo. Though not actually compiled by Milne, it was due to his initiative that the register was begun and continued. The first part, from 14 Sep. 1872 to 5 Oct. 1875 (nos. 1–41), is copied from a list by E. Knipping. From 6 Oct. 1875 to the end of March 1885 (nos. 42–606), the records were made by a Palmieri seismograph. From 1 Apr. 1885, they were continued by a Gray-Milne seismograph, which Milne placed in the Meteorological Observatory at Tokyo. As this instrument was in part provided by a grant from the British Association, the Gray-Milne records were published in instalments in Milne's reports and circulars, the last entered being no. 2637 on 31 Dec. 1902. Thus, with the exception of an interval of about two months at the end of 1882, when the Palmieri seismograph was removed to another station, we have an instrumental record extending over more than 27 years, and containing the entries of 2596 earthquakes. The early records (565 in number) provided by the Palmieri seismograph give only the time of occurrence and the intensity and direction of the shock. When the Gray-Milne seismograph was installed, the duration of the shock and the maximum period in seconds and amplitude in millimetres of the horizontal motion were added, and, later, the same elements for the vertical motion. In its fullness and long duration, the Tokyo list thus possesses a value that can hardly be over-estimated*.

201. Distribution of Japanese Earthquakes. It was characteristic of Milne that he left the discussion of his great catalogue

* *Japan Seis. Soc. Trans.* vol. 2, 1880, pp. 4–14, 39; vol. 6, 1883, pp. 32–35; vol. 8, 1885, pp. 100–108; vol. 10, 1887, pp. 97–99; vol. 15, 1890, pp. 127–134; *Brit. Ass. Rep.* 1886, pp. 414–415; 1887, pp. 212–213; 1888, pp. 435–437; 1889, pp. 295–296; 1890, pp. 160–162; 1891, pp. 123–124; 1892, pp. 93–95; 1893, pp. 214–215; 1895, pp. 81–84, 113–115; 1897, pp. 132–137; 1898, pp. 189–191; 1899, pp. 189–191; *Brit. Ass. Seis. Com. Circulars*, no. 1, pp. 29–30; no. 3, pp. 90–92; no. 5, pp. 142–144; no. 7, pp. 223–225.

for the most part to other workers*. He was mainly interested himself in studying the distribution of the Japanese earthquakes in space. For this purpose, he divided the map of the whole country into more than 2000 rectangles by north-south and east-west lines one-sixth of a degree of longitude and latitude apart. These rectangles are all numbered, and the position of the epicentre in the catalogue is indicated by the number of the rectangle in which it lies. On the map, the epicentres are represented by dots spread uniformly over each rectangle, and it is seen at once that these dots are grouped in 15 districts with definite boundaries, except that, in two districts, the shocks were so numerous that the boundaries are represented as straight lines beyond the proper limits, as there would not otherwise have been room to insert the dots.

The map confirms in detail the laws of distribution which Milne had already deduced from the earthquakes of north Japan during the years 1881–83. (i) Of those earthquakes, 84 per cent. originated near the sea-coast or beneath the ocean. He again notices that the majority of earthquakes originate along the east coast and that in many of them the epicentres are submarine. Drawing lines in east and south-east directions from the highlands of Japan into the Pacific Ocean, he found that the slope in places is as much as 1 in 30 or 1 in 20, and that it was in such districts, where the surface-flexures are greatest, that the earthquakes are most frequent and violent. (ii) A second and somewhat unexpected result is that the central portion of Japan, in which active volcanoes are numerous, is singularly free from earthquakes†. (iii) Milne showed that, with two or three exceptions, the districts in which earthquakes are frequent are those in which movements of secular elevation and depression are taking place. (iv) Lastly, a map drawn by Milne, though not published, shows the distribution of earthquakes accompanied by sound. Generally, it appears, sound is heard in rocky mountainous districts; on alluvial plains, it is but rarely observed.

* The following references may be given to papers founded on Milne's catalogue: *Roy. Soc. Proc.* vol. 60, 1897, pp. 457–466; *Geol. Soc. Quart. Jl.* vol. 53, 1897, pp. 1–15; *Geogr. Jl.* vol. 10, 1897, pp. 530–535; *Geol. Mag.* vol. 4, 1897, pp. 23–27.

† See *Geogr. Jl.* vol. 10, 1897, p. 534.

202. Catalogue of Destructive Earthquakes. The homogeneity of the second catalogue of Japanese earthquakes fails us in catalogues extending over many centuries. Milne sought to restore this valuable feature to his last great work—the catalogue of destructive earthquakes—by confining it to shocks which have caused some marked injury to buildings, those due to "movements which have probably resulted in the creation or extension of a line of fault." Even so, the catalogue is inevitably incomplete. For instance, during the first thousand years of the Christian era, the records of three out of every four earthquakes come from Italy, China and Japan; during the nineteenth century, the ratio drops to one in five. Many of our seismic records only began to exist during the last five centuries, that of America in 1520, of the Philippine Islands in 1600, of India in 1668, and of New Zealand in 1848. Nevertheless, unless some unexpected sources of information should be disclosed, it seems unlikely that very material additions can be made to Milne's catalogue. It must be regarded as exhaustive as it is now possible to render it.

The catalogue, which occupied Milne for several years, was published in 1911. It covers a period of 1893 years, from A.D. 7 to 1899, and contains 4151 entries. As far as possible, it states for every earthquake the year, month and day of its occurrence; the country visited by it, and often the part of the country in which it was felt most strongly; the intensity of the shock according to a scale of three degrees; and, lastly, the authorities for the information given. Sometimes, a few details are added, such as the number of lives lost or of houses destroyed, which furnish a more precise idea of the disastrous character of the earthquake*.

203. Distribution of Earthquakes in Time. Incomplete as the catalogue may be for the whole world, Milne points out that, for Europe from the year 1000, the most destructive of all earthquakes can hardly have escaped record. They were events of

* *Brit. Ass. Rep.* 1911, pp. 649–740. Among the papers founded on this catalogue, references may be given to the following by F. de Montessus de Ballore: Paris, *Ac. Sci. C. R.* vol. 154, 1912, pp. 1843–1844; vol. 155, 1912, pp. 560–561; vol. 156, 1913, pp. 100–102, 414–415, 1194–1195.

historical importance. Grouping them in periods of fifty years, from A.D. 1000 to 1850, he shows that, from about the year 1650, there has been a rapid increase in frequency, the numbers in successive half-centuries from 1600 being 3, 10, 15, 17 and 30. He suggests that these figures indicate a real increase in seismic activity from about 1650, accompanied, as he showed later, by a corresponding increase in volcanic activity.

The rest of Milne's work on the variations of earthquake-frequency must be noticed somewhat slightly, as, owing to the brevity of his record of world-shaking earthquakes, the results must be regarded as suggestive rather than as established. Like Mallet, he was attracted by the alternations of seismic activity and repose, and he showed that earthquakes have a tendency to occur in groups, the number in a group usually varying from two to fifteen or more, the groups lasting from one to three days, seldom more than six days. The intervals between the centres of successive groups vary from 15 to 50 days, being roughly proportional to the intensity of the earlier group.

More remarkable still is the tendency of strong earthquakes to occur in pairs or even in triplets, in districts so widely separated as, for instance, Guatemala and the Indian Ocean (19 Apr. 1902), the interval between them being in this case less than the time required for the first preliminary tremors to cross the distance between the foci. Milne records 15 cases of great double earthquakes, and two of triple earthquakes, during the years 1899–1906.

Such inquiries lead naturally to the suggestion that the variations in frequency in widely separated districts may to a certain extent be synchronous. The materials are perhaps at present insufficient for such wide generalisations, but Milne gives reasons for supposing that, in two districts so far apart as the eastern and western margins of the Pacific, the frequency in one increases and decreases simultaneously with that in the other; and that, during the last two centuries, Italy, Japan, China and America have roughly agreed in their periods of activity and repose.

The last point to which I can refer is Milne's interesting suggestion that the frequency of great earthquakes may be connected with the migrations of the pole. The earthquakes of the years 1892–1904 were grouped in successive intervals of 36½ days each. Counting the earthquakes in those intervals in which changes of direction took place in the movements of the pole, and in the preceding and following intervals, Milne found the numbers to be respectively 167, 287 and 217, and he thus concludes that great earthquakes are frequent about the times when changes occur in the direction of the polar movements, and especially when those changes occur most rapidly*.

204. Study of Distant Earthquakes. In this section, I propose to touch very briefly on Milne's contribution to this fascinating department of seismology.

As far back as 1883, he had foreseen the new development when he wrote that "it is not unlikely that every large earthquake might, with proper instrumental appliances, be recorded at any point of the land-surface of our globe." But it was not until 1889 that his attention was thoroughly aroused. In that year, many earthquakes of distant origin, such as those of 17 Apr. and 28 July in Japan, were registered by horizontal pendulums at Potsdam and Wilhelmshaven—instruments that had been erected there by Rebeur-Paschwitz (1861–95) in order to detect and measure the lunar disturbance of gravity.

205. Few men have worked so bravely as Ernst von Rebeur-Paschwitz. He was born on 9 Aug. 1861, and obtained his doctor's degree at the university of Berlin in 1883, becoming an assistant first in the observatory of Berlin and then in that of Karlsruhe. In 1884, his attention was directed to Zöllner's horizontal pendulum, and it was a form of this instrument modified by him that proved to be so sensitive to the vibrations of distant earthquakes†. It was about this time that the first symptoms of a fatal illness developed, and, during the last years of his life, his work was carried on in continual pain and sickness. Yet, as Milne remarks, "it was during this period of physical incapacity that von Rebeur produced his most remarkable work, and became the pioneer of a new seismology." He died on 1 Oct. 1895.

* *Brit. Ass. Rep.* 1900, pp. 107–108; 1903, pp. 78–80; 1906, pp. 97–99.
† *Brit. Ass. Rep.* 1893, pp. 303–306.

In many ways, Rebeur-Paschwitz seemed to resemble our incarnation of the ideal man of science. He had Darwin's lovable nature, as well as his modesty and utter carelessness of his own fame. But the likeness was closest in the unceasing energy with which he laboured, in spite of the constant suffering that would have made many stronger men feel that their life's work was ended. Dying at the age of 34, he had achieved what others of twice the age might regard with satisfaction as the fruits of a well-spent life*.

206. Stimulated by Rebeur-Paschwitz' observations, we find Milne recording distant earthquakes in Tokyo in 1893 with his own seismograph, and his last report (1895) on the earthquake-phenomena of Japan contains a list of some of the shocks recorded there before his observatory was destroyed. He reached Shide in the Isle of Wight on 30 July 1895, and, within three weeks, a brick pier was built and the first horizontal pendulum erected in what afterwards became the principal earthquake-observatory in the world.

In 1894, Rebeur-Paschwitz had suggested the foundation of an international system of earthquake-observatories, a plan cut short by his early death. Shortly afterwards, Milne conceived independently a similar and much wider network of stations, and, in 1897, the Seismological Committee adopted his horizontal pendulum as the standard instrument for the survey. Beginning with his observatory at Shide, the network of stations extended year by year, until the number of stations contributing records to the Committee amounted to 34. In the British possessions, they were to be found in Canada and British Columbia, in Ascension and the Cape of Good Hope, and in various districts of India, Australia and New Zealand. Records were also sent to the Committee from several observatories in foreign countries—from Spain, the Azores and Syria, and from such distant island stations as Fernando Noronha and Honolulu.

* *Nature*, vol. 52, 1895, pp. 599–600; *Geol. Mag.* vol. 2, 1895, pp. 575–576; *Ital. Socs. Sism. Boll.* vol. 1, 1895, p. 137. For Rebeur-Paschwitz' principal memoirs, see *Ac. Nat. Curios. Nova Acta*, vol. 60, 1892, pp. 1–216, and *Beitr. Geophys.* vol 2, 1895, pp. 211–536; and, for a summary of his earlier work, *Brit. Ass. Rep.* 1893, pp. 309–334.

The reports of the Seismological Committee, so far as Milne was responsible for them, are chiefly concerned with the discussion of the registers from these various stations. He touched, however, on many subjects, some of which—such as earth-tremors, the diurnal wave, and the tilting due to tidal loading—have no necessary connexion with earthquakes. The sections on the distribution of earthquakes in time have been already noticed (art. 203). Some others relating to distant earthquakes—such as earthquake-echoes, the nature of the large waves, and the dissipation of earthquake-motion as the waves radiate outwards—must here be passed over. In the remaining outline of Milne's work, I will confine myself to two important subjects—the time-curves of the various phases and the distribution of the earthquake-origins in space.

207. (i) It is interesting to trace Milne's successive attempts to draw the average time-curves of the earthquake-motion*. The first, which is really a velocity curve, appeared in the report for 1897, but the observations so far obtained were discordant, and the only result deduced was an increase in the velocity of the early tremors with the distance, whether measured along the arc or the chord. In the following year, an advance was made by separating the preliminary tremors and the large waves, and a curve was given showing very clearly that the average duration of the preliminary tremors increased with the arcual distance from the origin. In 1900—these years are dates in the history of seismology—the first time-curves proper were drawn. Distances from the epicentre in degrees of arc were measured along the horizontal axis, and the times of transit of the initial preliminary tremors and the maxima of the large waves to particular stations in the perpendicular direction. The time-curve for the large waves is practically a straight line, corresponding to a constant arcual velocity of 3 km. per second; that for the preliminary tremors shows that the velocity along the arc increases with the distance. Two years later, the curves begin to assume their now familiar form; and, as they are based on records from Milne seismographs

* *Brit. Ass. Rep.* 1897, pp. 171–176; 1898, pp. 220–224; 1900, pp. 65–67; 1902, pp. 64–67; 1903, pp. 84–85.

only, the divergences from the average are less noticeable than before. It had been shown by R. D. Oldham in 1900* that the preliminary tremors are divisible into two series, the first con-

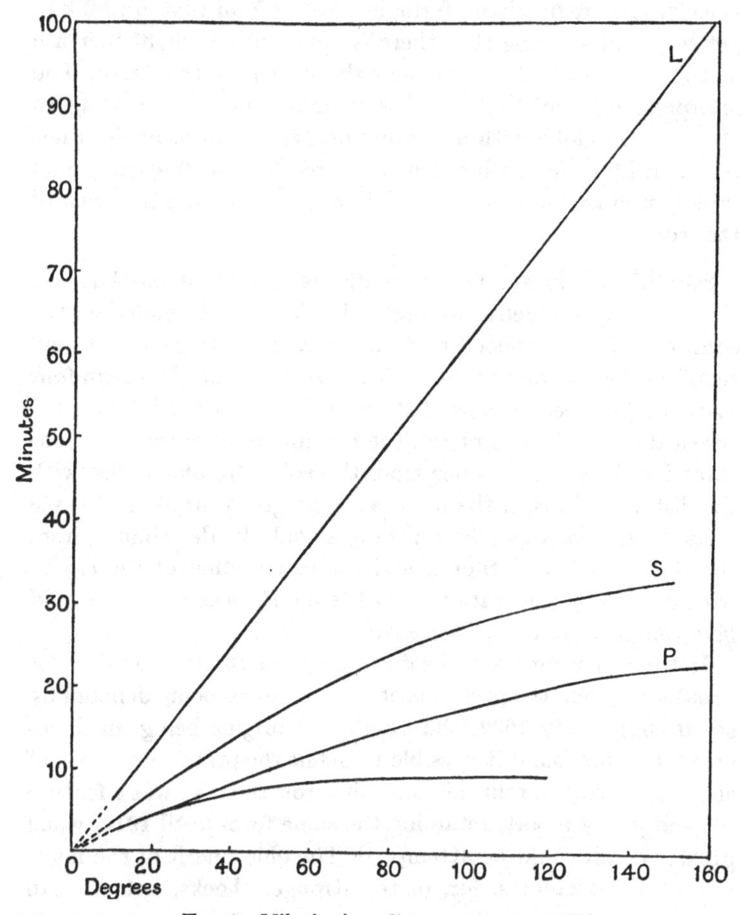

Fig. 9. Milne's time-distance curves (1902).

sisting mainly of condensational, the second of distortional, vibrations. A curve of the duration of the first preliminary tremors was also added (Fig. 9). In the 1903 report, the bearing

* *Phil Trans.* 1900, A, pp. 135–174.

of these curves is discussed. Measured along the arc from 20° to 150° from the origin, the velocity of the first tremors increases from 10·5 to 16·3 km. per sec. Measured along the chord, the velocity is more nearly uniform, increasing from 10·5 to 12·0 km. per sec., and showing that there is apparently a slight increase in speed as the paths approach the centre of the earth. The interpretation which these observations require, according to Milne, is a "globe with an approximately uniform nucleus not less than $\frac{19}{20}$ of the earth's radius covered by a shell which passes rapidly upwards into the materials which constitute the crust of the world."*

208. (ii) To locate the epicentre of a distant earthquake, Milne employed chiefly two methods. The first depended on the constancy of the velocity of the maximum large waves and required the absolute times of their passage at not less than four stations. The second depended on the fact that the interval between the initial tremors and the maximum large waves is constant for the same distance from the epicentre and varies with the distance. Thus, if the intervals were given for at least three stations, the corresponding distances could be determined from the time-curves and tables, and the intersection of the circles with centres at the stations and these distances as radii would give the position of the epicentre.

In the report for 1900, the first map of earthquake-origins (for 1899) was given, the positions of the epicentres being denoted by small circles. By 1902, the number of origins being much increased, Milne found it possible to define the principal regions of seismic activity. From this time onwards, the map was a feature of each year's report, retaining the same form until 1910, when greater detail could be attempted. The chief merit of the maps is their completeness, for, of the stronger shocks, not one can

* Since 1902, many hundreds of observations have been collected, and these have naturally modified Milne's curves and the tables founded on them. The latest and most trustworthy tables are those prepared by H. H. Turner (1861–) and published in the Circulars issued by the Seismological Committee. The bearing of the time-curves, and especially of those of the second preliminary tremors on the constitution of the earth has been discussed by several writers, in this country by R. D. Oldham (*Geol. Soc. Quart. Jl.* vol. 62, 1906, pp. 456–473) and C. G. Knott (art. 190, *Edin. Roy. Soc. Proc.* vol. 39, 1919, pp. 157–208).

now escape being recorded. On the other hand, the period that has elapsed since 1899 is but a brief interval in the history of the

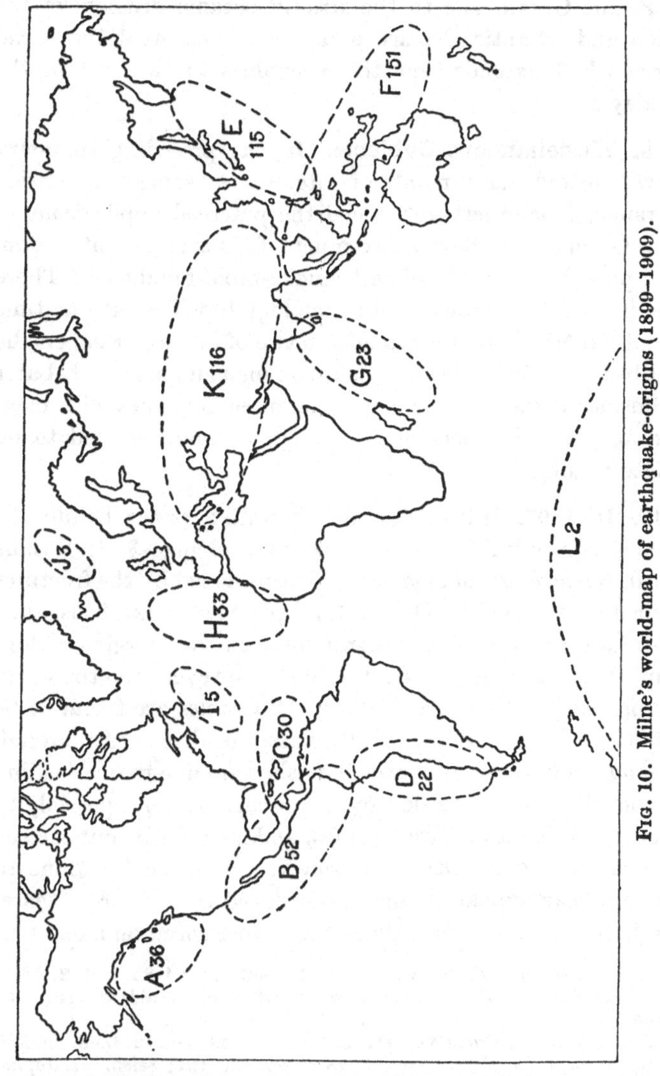

FIG. 10. Milne's world-map of earthquake-origins (1899–1909).

earth, and all that the maps can provide us with is an outline of those districts in which crustal changes are now taking place most rapidly. From the map for 1899–1909 (Fig. 10), we see that of every ten earthquakes, seven belong to the districts bordering the Pacific Ocean, one to the strictly oceanic districts of the Indian and Atlantic Oceans, and two to the great terrestrial district which extends from the Apennines to the east of the Himalayas.

209. Miscellaneous Subjects. In the preceding summary I have omitted many points of interest. As somewhat outside our range, I have left untouched the practical applications to which he humanely devoted so much of his time and attention, and especially the study of earthquake-proof buildings*. There were, indeed, few branches of seismology to which, at one time or another, Milne did not attend. In two of his earlier papers, he described the effects of earthquakes on men and animals†. Later he considered the possible connexion of earthquakes with magnetic and electric phenomena, though his inquiries led to no definite result‡.

210. In 1897, Milne read an interesting paper before the Royal Geographical Society on suboceanic changes§. In this, he turned to a new line of evidence—that provided by the fractures in our deep-sea cables. On the level plains of ocean-beds, such cables lie for years without harm; but, on the margins of suboceanic banks and of continental slopes, however deep the water may be, interruptions are frequent, and cables are found to be bent, twisted, or buried beneath masses of displaced material. In many cases, these submarine landslips are due to overloading by sedimentation, to erosion by ocean-currents, and possibly to the outbreak of fresh-water springs, and in so far lie outside the range of this book. Others, however, are traced by Milne to submarine earthquakes, which he believed to be more intense, as well as more numerous, than those experienced on land. Out

* *Japan Seis. Soc. Trans.* vol. 1, pt. 2, 1880, pp. 64–85; vol. 2, 1880, pp. 27–38; vol. 11, 1887, pp. 115–174; vol. 14, 1889, pp. 1–246; vol. 15, 1890, pp. 163–169.

† *Japan Seis. Soc. Trans.* vol. 11, 1887, pp. 91–111; vol. 12, 1888, pp. 1–4.

‡ *Japan Seis. Soc. Trans.* vol. 15, 1890, pp. 135–161; *Seism. Jl. Japan*, vol. 3, 1894, pp. 23–33.

§ *Geogr. Jl.* vol. 10, 1897, pp. 129–146, 259–285; *Brit. Ass. Rep.* 1897, pp. 130–206; 1898, pp. 251–254.

of 245 cable-fractures registered from 1872 onwards, Milne indicates 87 which occurred when delicate seismographs were in action. Of these, 58 coincided approximately in time with the disturbances registered by horizontal pendulums, while, in addition, 24 fractures occurred at the times of earthquakes felt on land, two of which were destructive along the neighbouring shores.

211. On only one important subject—the origin of earthquakes—did Milne touch less than might have been expected. His two popular volumes, *Earthquakes* and *Seismology*, summarise the views generally held at the times of writing*; and the references that he made to it here and there in his papers show that he favoured the theory that attributes earthquakes to the formation and growth of faults. In his early reports to the British Association, Milne states his views with a definiteness unusual for the time. From seismograph records and personal feelings, he traces "a sliding, jolting kind of motion, as might be produced by one mass of rock slipping over another." An earthquake so formed might consist of distortional vibrations only. Again, we have reason, he says, to believe that certain earthquakes "are the result of a sudden breaking in the rocky crust of the earth, produced by bending due, for example, to elevatory pressure."†

212. Conclusion. To state, as I have done for others, what seem to me to be the contributions of permanent value which Milne made to seismology would be merely to abbreviate and repeat the preceding pages. I will therefore confine these concluding remarks to two general points.

In the history of seismology, Milne's work must for all time occupy a large and worthy place. In our textbooks of the future, in books which aim at presenting a summary of existing knowledge, the references may be fewer as time goes on than his work deserves. It must be remembered that Milne was a student of earthquake-phenomena, and that our knowledge of the phenomena is always growing. A careful investigation of a great earthquake, on the other hand, retains its value for it can hardly be replaced.

* *Earthquakes* (1886), pp. 277–296; *Seismology* (1898), pp. 24–38.
† *Brit. Ass. Rep.* 1881, p. 201; 1883, p. 215.

Moreover, he preferred sometimes to start an inquiry and to leave others to finish it, to open up large views and to provide promising subjects of investigation for a future of wider knowledge. Of his two great catalogues—of Japanese earthquakes and of destructive earthquakes—his own analysis was slight. He was, in fact, content to provide the materials which others were to use in building.

The influence of a leader in science extends far beyond his own contributions. In founding the Seismological Society of Japan, which led to the formation of the Imperial Earthquake Investigation Committee, and in organising the network of stations throughout the world, and, above all, in his personal influence on those with whom he came in touch, it seems to me possible that seismology may owe almost as much to his guidance and inspiration as it does to the incessant labour of his well-spent life.

CHAPTER XI

FUSAKICHI OMORI

213. The interest in earthquakes that now prevails in Japan is a direct result of the work of the early British teachers in the Imperial University and especially of John Milne*. As already mentioned, the first organised investigation of Japanese earthquakes was due to his initiative. After 1883, as he gradually extended his "postcard barrier," the labour and expense grew beyond the portion of one man, and, at his request, they were taken over by the Imperial Meteorological Department with its numerous stations scattered over the empire. Up to this point, then, the government organisation was the lineal successor of that started by Milne. The next step, a natural one, was taken by a Japanese professor, and resulted in the creation of the Imperial Earthquake Investigation Committee.

THE IMPERIAL EARTHQUAKE INVESTIGATION COMMITTEE

214. Dairoku Kikuchi (1855–1917: created Baron Kikuchi in 1902) was born on 17 Mar. 1855 at Tokyo. For most of his education, he was indebted to English influences, for the years 1867–68 and 1871–73 were spent at University College School, London, and 1873–77 at St John's College, Cambridge. In the mathematical tripos of 1877, he was nineteenth wrangler. He then returned to Japan, where he lived a busy official life, serving as professor of mathematics in the Imperial University, Tokyo (1877–98), president of the University (1898–1901), Minister of Education (1901–03), president of the Imperial University of Kyoto (1908), and president of the Imperial Academy of Tokyo (1909). He died on 19 Aug. 1917, a "modest, courteous, gracious man," "always acting from the highest motives, strong in purpose yet never aggressive, and com-

* D. Kikuchi, Japan, *Earthq. Inv. Com. For. Publ.* no. 19, 1904, p. 2.

bining in a singular degree the finest traits of the Japanese Samurai with the best qualities of the youth of England."*

On 28 Oct. 1891, one of the greatest of Japanese earthquakes devastated large tracts in the provinces of Mino and Owari. The earthquake is a notable one on several accounts. It provided most of the materials for Omori's valuable studies of after-shocks. It revealed, though not for the first time, how complicated may be the distortions of the earth's crust during a great earthquake, and it led to the foundation of the Imperial Earthquake Investigation Committee. Omori's work will be referred to later in this chapter (arts. 223–241), but a short digression must be made here on Koto's investigation of the epicentral area.

215. Bunjiro Koto (1856–), professor of geology in the Imperial University of Tokyo, is perhaps more interested in the volcanoes, than in the earthquakes, of Japan. He has, however, written papers in Japanese on the geological aspects of two earthquakes, and his memoir (in English) on the cause of the Mino-Owari earthquake is one of the classics of seismology†. In this memoir, Koto traced the course of a remarkable fault that traversed valley, plain or mountain for a distance of forty miles, and, though he was prevented by the winter season from surveying it farther, that probably continued for another thirty miles as fas as Fukui. He showed that the ground on the north-east side of the fault was shifted to the north-west relatively to the other side through a distance of one or two metres, and that, except at one place, the north-east side is also depressed. When the vertical displacement was small, the soft earth on the surface was moulded into a rounded ridge, from one to two feet high, so that it resembled the path of a gigantic mole. When, however, the throw was great—and at Midori it amounted to 18 or 20 feet—the fault-scarp from a distance looked like a railway embankment. Described in simple terms, Koto's memoir brought conviction, where doubt had hitherto prevailed, as to the dislocation of the crust in a great earthquake‡.

* *Nature*, vol. 100, 1917, pp. 227–228; *Who's Who in Japan*, 1912, pp. 362–363.

† Tokyo, *Coll. Sci. Jl.* vol. 5, 1893, pp. 295–353.

‡ Other important memoirs on this earthquake are those by A. Tanakadate (1856–) and H. Nagaoka (1865–), Tokyo, *Coll. Sci. Jl.* vol. 5, 1893, pp. 149–192, and J. Milne, *Seism. Jl. Japan*, vol. 1, 1893, pp. 127–151.

216. In December 1891, Kikuchi, supported by a great majority in the House of Peers, carried a petition to the Government to appoint a Committee for the study of earthquakes. Six months later, on 25 June 1892, an Imperial Ordinance decreed the foundation of "a Committee for investigating the prevention of earthquake disasters, more usually known as the Earthquake Investigation Committee."

This Committee was provided with a twofold object—(i) to ascertain whether there are any means of predicting earthquakes, and (ii) to investigate what can be done to reduce the disastrous effects of earthquakes to a minimum, by the choice of proper structures, materials, position, etc. From the first, the Committee has taken a wide view of its office, and, while always keeping the above objects in view, it has not hesitated to undertake any inquiries that seemed likely to add to our knowledge, even if their practical bearing were not immediately obvious.

It was decided that the Committee should consist of a president, a general secretary, and (normally) 25 members. They were to be chosen from among seismologists, physicists, geologists, civil engineers, architects, etc.*, and it was held—surely a wise, if unusual, provision—that "a high position must be assured to members of the Committee," while the president must be an official of the rank of Tcho-kunin.

The first president was Hiroyuki Kato, afterwards Baron Kato (1836–), president of the Imperial University of Tokyo and one of the pioneers in the introduction of western learning into Japan. Kikuchi was naturally the first secretary. In 1893, he succeeded Kato as president and held the office until 1901, being followed by K. Tatsuno (1854–), B. Mano (1861–), etc. S. Sekiya was secretary from 1893 to 1896, and F. Omori from 1897 to 1923, when he was succeeded by A. Imamura.

217. How active the Committee has been throughout is evident from the volume of its proceedings. It began in 1893 with the issue of *Publications in the Japanese Language*. Of these,

* In 1899, out of a total of 28 members, at least twelve were engineers and three architects, the rest including the professors of seismology, geology, mathematics and physics in the Imperial University of Tokyo. Of these 28 members, no less than 23 have contributed papers to the publications issued by the Committee.

98 parts had appeared by the close of 1922, the total number of papers in them being 312, and of pages 8836. Many of the papers of general interest were translated into English and printed, a few of them in the *Journal* of the College of Science, Imperial University, Tokyo, but most in a new series of *Publications in Foreign Languages*. These were begun in 1897, the last, no. 26, being published in 1908*. The number of papers in the foreign section is 56, containing 2260 pages. As the separate articles were of considerable length (40 pages on an average), a new journal, the *Bulletin*, was started in 1907, in order "to secure a quick publication of short notes and preliminary reports on seismological subjects." Of this, eight complete volumes have been published and parts of three others, the last in 1922. With the suspension of the *Publications*, however, the length of the papers in the *Bulletin* rapidly increased, until we have the sixth and seventh volumes, of 713 pages, occupied by Omori's memoirs on the eruptions and earthquakes of the Asamayama in 1911–1917, and the eighth, of 525 pages, with the memoirs on those of the Sakura-jima in 1914. In consequence, a new journal called *Seismological Notes* was begun in 1921†. Of this, six numbers of 119 pages had been issued by the close of 1924. Thus, the total number of pages in the four series down to December 1924 is no less than 13,453‡.

218. Many of the papers published in Japanese only are of local value. Such, for instance, are the reports on the damage to buildings caused by the earthquakes of Mino-Owari in 1891, Hokkaido and Shonai in 1894, Tokyo in 1894 and 1895, San-Riku in 1896, Sendai in 1897, Osaka in 1899, Sakura-jima in 1914, and, outside Japan, of Assam in 1897. Of more general interest are the rules for the construction of earthquake-proof buildings, in wood or otherwise, and for the repair of damaged chimneys. It would be too much to expect that the 1247 pages

* No. 25 was in preparation in 1913, but, as far as I know, has not yet appeared.

† In the following pages, the *Publications in the Japanese Language*, the *Publications in Foreign Languages*, *Bulletin* and *Seismological Notes* will, for brevity, be referred to as *Jap. Publ.*, *Publ.*, *Bull.* and *Notes*, respectively.

‡ How lavishly these works are illustrated will be evident from the fact that the *Publications in the Japanese Language* are accompanied by 2299 plates, the *Publications in Foreign Languages* by 520, and the *Bulletin* by 444.

of materials for the earthquake-history of Japan should be translated into English. But a seismologist cannot help regretting that such papers as those of Omori on tsunamis and on Formosa earthquakes, of Iki on the San-Riku tsunamis of 1896, of Sano on the Formosa earthquake of 1904, or of Imamura on the earthquake-zones in the south-west of the Main Island, should remain locked up in what to most of us is a sealed language.

Welcome features in the Japanese series are the number of different authors, the number of papers written jointly by two or more authors, and the number of articles written by others than members of the Committee. If we were to judge by the *Publications in Foreign Languages* alone, we might be left with the impression that, up to 1923, there were only two seismologists in Japan, although about a dozen other workers might make an occasional contribution to the science. In the Japanese section of the *Publications*, there are no fewer than 82 authors, the most prolific being Omori, who contributes as sole author 98 articles containing more than 4000 pages.

Of the *Publications in Foreign Languages*, the first two numbers, on the variation of latitude, lie outside our range. The third relates chiefly to the organisation of the Committee, and is the only one not written in English. Of the 54 memoirs in the remaining numbers, 33 are the work of Omori, one a joint paper by Sekiya and Omori, and 20 by other writers. Again, in the eleven volumes of the *Bulletin*, 78 papers are by Omori, two by Imamura, and one by C. Darleth. The papers in English will thus be considered in connexion with their principal author, though a brief reference must be made to the important memoirs by some authors who are still alive, and especially to those by H. Nagaoka (1865–) on the elastic constants of rocks*, by S. Kusakabe on the modulus of elasticity of various rocks†, by A. Imamura on seismograms furnished by Milne horizontal pendulums at Tokyo‡, and by K. Honda, T. Terada, Y. Yoshida and D. Ishitani on the secondary undulations of oceanic tides§.

* *Publ.* no. 4, 1900, pp. 47–67.

† *Publ.* no. 14, 1903, pp. 1–73; no. 17, 1904, pp. 1–48; and no. 22 B, 1906, pp. 27–49.

‡ *Publ.* no. 16, 1904, pp. 1–117.

§ *Publ.* no. 26, 1908, pp. 1–113.

SEIKEI SEKIYA

219. Even if he were not the author of some valuable memoirs on earthquakes, Seikei Sekiya (1855–96) would require mention in these pages as the first professor of seismology in any university in the world. Prompted by Kikuchi, he took up the study of earthquakes in 1880 and was placed in charge of a small seismological laboratory in the Imperial University at Tokyo. Appointed to the newly-created chair of seismology in that university in 1886, he was the natural successor to Kikuchi as secretary to the Imperial Earthquake Investigation Committee. By his influence and persuasive power, by his kindly disposition and straightforward manner, he helped greatly in the extension of the seismic survey of Japan and in the erection of seismographs throughout the country. In 1885 the number of observing stations was over 600. At the time of his death (9 Jan. 1896), it had risen to 968*.

220. Beginning his scientific career about the time when accurate seismographs were invented, it was only natural that some of Sekiya's most valuable work should consist in the measurement of their records. One of his earliest papers was that on the severe earthquake of 15 Jan. 1887, the epicentre of which was 35 miles S.W. of Tokyo. The maximum horizontal motion (double amplitude), as recorded by his seismographs, was 21 mm., the maximum vertical motion 1·8 mm., and the maximum acceleration 66 mm. per sec. per sec.—one of the earliest estimates ever made of the latter element from precise data. In his examination of the epicentral area, he notices the slightness of the damage to wooden houses; but in Japan, he adds, "fire is a more constant and even more dread enemy than earthquakes."†

Outside the ranks of specialists, Sekiya is most widely known by his model showing the motion of an earth-particle during this earthquake‡. By this happy device, he has given the best illus-

* *Publ.* no. 19, 1904, p. 4; *Geol. Mag.* vol. 3, 1896, pp. 143–144; *Nat. Sci.* vol. 8. 1896, p. 279.

† *Japan Seis. Soc. Trans.* vol. 11, 1887, pp. 79–88; *Nature*, vol. 36, 1887, pp. 379–381.

‡ *Japan Seis. Soc. Trans.* vol. 11, 1887, pp. 175–177; *Nature*, vol. 37, 1888, p. 297.

tration that we possess of the complicated movements of the ground during a severe earthquake.

Sekiya was one of the first to attend closely to the vertical motion during earthquakes. During two years (1885–87), he recorded 100 earthquakes, excluding tremors, at Tokyo. He found that vertical motion was present in only 28 of these seismograms, that the amplitude of the maximum vertical motion was on an average about one-sixth of the maximum horizontal motion, its period about one-half and its duration about one-third*.

In conjunction with F. Omori, Sekiya wrote two interesting papers. In the first, they described some experiments suggested by an earlier work of Milne's (art. 198). They compared the records of thirty earthquakes made by two similar seismographs, one placed at the bottom of a pit 18 feet deep, the other on the surface. In slight earthquakes and in the large undulations of severe earthquakes, the amplitude and maximum acceleration were nearly the same in both positions; but, in the ripples of the severe earthquakes, these elements were respectively two and five times as great on the surface as at the bottom of the pit†. In the second paper, they reproduced the first "clear instrumental record of a destructive earthquake" ever taken in Japan or elsewhere—that of the Tokyo earthquake of 20 June 1894. The record was remarkable for the extreme simplicity of the principal movement, the range of which was 73 mm. in the direction S. 70° W. The mean direction of overthrow of 245 stone-lanterns in Tokyo was found to be S. 71° W.‡

For two or three years before his death, Sekiya, in spite of continued ill-health, was engaged in compiling the Earthquake Investigation Committee Catalogue of Japanese earthquakes§, issued and edited by his successor, F. Omori, in 1899. Founded on records contained in 427 different Japanese chronicles, monographs and unpublished journals, this valuable catalogue gives the date, intensity and district of 1898 earthquakes during

* *Japan Seis. Soc. Trans.* vol. 12, 1888, pp. 83–106.
† *Japan Seis. Soc. Trans.* vol. 16, 1892, pp. 12–45.
‡ Tokyo, *Coll. Sci. Jl.* vol. 7, pt. 5, 1894, pp. 1–4.
§ Tokyo, *Coll. Sci. Jl.* vol. 11, 1899, pp. 315–388.

the 1451 years from A.D. 416 to 1867*. In connexion with this catalogue, should be mentioned the "Materials for the Earthquake History of Japan from the earliest times down to 1866," compiled under the direction of S. Sekiya and F. Omori†. This is a great work of more than 1200 pages, written in Japanese. The appendix to the second part consists of a catalogue of earthquakes, presumably the same as that referred to above.

FUSAKICHI OMORI

221. How great was the part taken by Omori in the work of the Investigation Committee has been mentioned above. He wrote more than half of the *Publications in the Japanese Language*, nearly three-quarters of the *Publications in Foreign Languages*, and 98 per cent. of the *Bulletin*. Altogether, in the four different journals, his contributions amounted to just twice the total written by his colleagues.

Fusakichi Omori was born at Fukui on 30 Oct. 1868. In 1886 he entered the College of Science in the Imperial University of Tokyo, taking the course in physics. After his degree (1890), he took post-graduate courses in seismology and meteorology. John Milne was then professor of geology in the College of Engineering and Seikei Sekiya professor of seismology in the College of Science. Encouraged by Milne, he studied the after-shocks of earthquakes, and his memoir on this subject (1894) was the first attempt to examine with precision their distribution in time and space. In 1891 he had been appointed assistant to Sekiya, in 1892 a member of the Imperial Earthquake Investigation Committee, and in 1893 lecturer on seismology in the university. He wrote one or two papers in conjunction with Milne and also with Sekiya. In 1895, Omori was sent to Italy and Germany for two years' further study. During his absence, his old teacher Sekiya died on 9 Jan. 1896, and, in 1897, Omori succeeded him as professor of seismology and also as secretary of the Earthquake

* Catalogues of Japanese earthquakes had previously been compiled by I. Hattori, E. Naumann and J. Milne (arts. 185, 199) and by H. Okajima in his *History of unusual natural phenomena in Japan* (1894).

† *Jap. Publ.* no. 46, 1904, pt. 1, pp. 1–606; pt. 2, pp. 1–596; Appendix to pt. 2, pp. 1–45.

Investigation Committee. Twenty years later, he became acting president of the Committee. But, whether as secretary or president, he was the natural leader of earthquake studies in Japan, and, during the last 28 years of his life, there flowed from his pen a continual stream of memoirs, written with few exceptions in Japanese or English, many of those in the former language being also translated into the other.

Besides the earthquakes of his own country, Omori investigated several in other lands. He visited the epicentral areas of the Kangra earthquake of 1905, the Californian earthquake of 1906, and the Messina earthquake of 1908. As Japanese delegate, he frequently attended meetings of international associations. In the third quarter of 1923, he visited Australia for the meeting of the Pan-Pacific Science Congress. On the journey home his health steadily declined. He reached Tokyo to find it ruined by the great earthquake and fire of 1 Sep., and died on 8 Nov. in the university hospital, not far from the damaged Seismological Institute, where he had spent the greater part of his active life, and which he had made known throughout the world as a centre of earthquake-investigation. A day or two before his death, he received from the Imperial Court the highest order of the Sacred Treasure, the greatest honour, I believe, that can be conferred on any Japanese subject. But a still higher honour, it seems to me, was the regret felt in every country in the world at the loss of a kindly courteous gentleman and of one whose advice and experience would have been invaluable in the study of the remarkable earthquake of Sagami Bay*.

222. Omori's work, like Milne's, covered the whole range of seismology. In the attempt to describe it, I have grouped his papers in sections, which are arranged as a rule in the chronological order of the most important paper in each. Thus, while his earliest papers were jointly written, his first important memoir was that on after-shocks. This and later papers on the same subject may be conveniently grouped together.

* For some of the above details I am indebted to a brief notice in *Who's Who in Japan*, 1912, p. 691, and to a memorial note by Profs. Imamura and Yamazaki.

223. After-Shocks of Earthquakes. Omori was fortunate in possessing the first available series of after-shocks of great earthquakes—of the Mino-Owari earthquake of 28 Oct. 1891, the Kumamoto earthquake of 28 July 1889, and the Kagoshima earthquake of 7 Sep. 1893. The areas disturbed by these earthquakes contained 321,000, 39,000 and 30,000 square miles, respectively, and the numbers of after-shocks during the first thirty days were 1746 at Gifu, 340 at Kumamoto, and 278 at Kagoshima, thus suggesting that the frequency of after-shocks increases with the magnitude of the disturbed area.

Of the 3365 after-shocks recorded during the first two years at Gifu, 10 were violent, 97 strong, 1808 weak, 1041 feeble, and 409 merely earth-sounds. Nine of the violent shocks were felt during the first four months, all the strong shocks within the first thirteen months, and all the weak shocks within the first twenty months.

In each earthquake, the rapid decline in frequency was marked. Omori found that the formula

$$y = \frac{k}{x + h}$$

agreed closely with the observed numbers, y being the number within a given interval distant x from the initial epoch, and h and k constants. For the Gifu series, taking the ten half-daily numbers of shocks from 29 Oct. to 2 Nov., and using the method of least squares, he found that

$$y = \frac{440{\cdot}7}{x + 2{\cdot}31},$$

a formula which represents with fair accuracy the frequency of after-shocks a year or two later, and even after the lapse of eight years. Omori's diagram showing the decline in monthly frequency of after-shocks at Gifu is reproduced in Fig. 11, the broken line representing the rectangular hyperbola determined from the monthly numbers.

The number of after-shocks recorded during the first two years diminished rapidly with increasing distance from the central point of the Neo Valley. It was 3365 at Gifu (17 miles), 1298 at Nagoya (37 miles), 70 at Osaka (88 miles), and 30 at Tokyo (166 miles). Indeed, Omori found himself able to represent the variation in the number of after-shocks by series of equal-frequency curves. He also discusses the periodicity of after-shocks, concluding that his curves seem to indicate a very slight diurnal variation of the mean frequency, the maximum occurring about 1 a.m.*

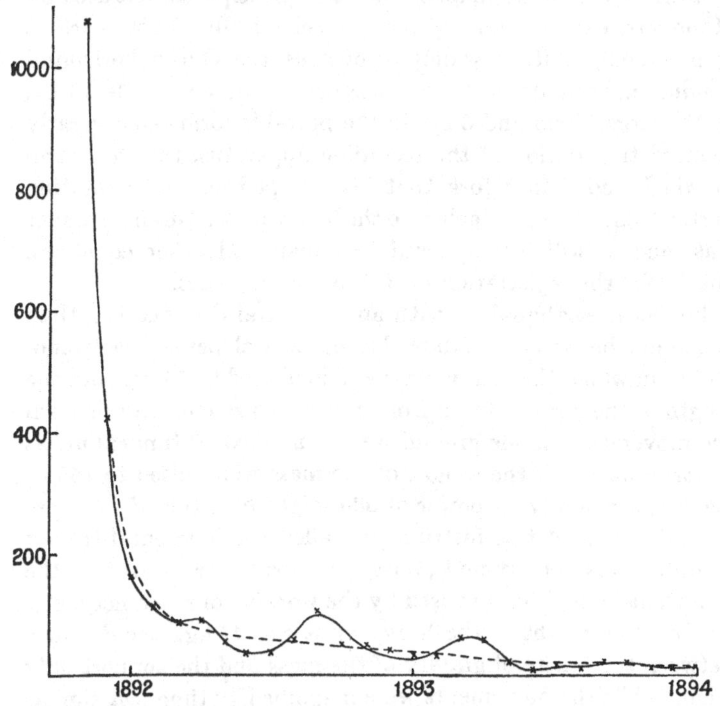

FIG. 11. Curve illustrating the decline in frequency of after-shocks at Gifu (Oct. 1891–Dec. 1893).

On several occasions, Omori returned to the subject of after-shocks, considering those of the great Japanese earthquake of 4 Nov. 1854, the Hokkaido earthquake of 22 Mar. 1894, and the earthquakes of 8 May 1847 near Matsushiro and of 9 Jan. 1848 near Kyoto. He shows that the earthquakes of 1847 and 1891 were followed by about the same numbers of after-shocks, and

* Tokyo, *Coll. Sci. Jl.* vol. 7, 1894, pp. 111–200; *Seism. Jl. Japan*, vol. 3, 1894, pp. 71–80; *Publ.* no. 7, 1902, pp. 27–51.

that the monthly after-shock frequency of the earthquake of 1848 was practically identical with that of the Mino-Owari earthquake after the lapse of forty days*.

224. Horizontal Pendulums. Omori devised several forms of horizontal pendulum, all on the same principle as that used by Milne (art. 196), but all registering mechanically. In the earliest form (1898), that so widely known as the Omori horizontal pendulum, he made the heavy mass comparatively small—14 kg. in the larger form and 3 kg. in the portable form—and greatly reduced the friction of the recording apparatus, the short arm of which ended in a fork that just clasped an easily rotating vertical axle of steel attached to the heavy mass. This instrument was, and is still, widely used in Japan and other countries, chiefly for the registration of distant earthquakes.

For local earthquakes, with an epicentral distance less than 1000 km., he designed (1902) his horizontal pendulum tromometer, in which the heavy mass was increased to 50 kg., and the length of the strut reduced from a metre to 20 cm. In this form the movements of the ground were magnified 20 times; but, as Omori remarks, if the weight of the mass were raised to 150 or 200 kg., a magnifying power of 330 might be attained. A more compact form of this instrument, called the horizontal tremor recorder, was constructed (1904) principally for measuring the vibrations in buildings caused by the working of machinery, etc. In this, the weight of the heavy mass was 15 kg., the distance between the centre of gravity of the mass and the support only 6 cm., while the movements were magnified 70 times. A similar instrument, with a heavy mass of 32 kg. and a magnifying power of 200, was made in 1907. Both tromometer and recorder were found of great service in the study of local shocks, and especially of those connected with the eruptions of the Usu-san and Asama-yama†.

225. Nature of Earthquake-Motion. For our knowledge of the elements of the earthquake-motion in local earthquakes, we are

* *Ital. Soc. Sism. Boll.* vol. 2, 1896, pp. 152–155; *Publ.* no. 4, 1900, pp. 39–45; *Bull.* vol. 2, 1908, pp. 185–195.

† Tokyo, *Coll. Sci. Jl.* vol. 11, 1899, pp. 121–145; *Publ.* no. 12, 1902, pp. 1–7; no. 18, 1904, pp. 1–3, 5–12; *Bull.* vol. 1, 1907, pp. 191–193.

indebted mainly to the seismologists of Japan—in the first place to Ewing, Milne and Sekiya, and later to Omori, who had at his disposal more varied instruments and far ampler materials. The most elaborate of his purely seismological papers are those which he has devoted to this subject. If collected together, they would form a volume of more than 800 pages.

Omori followed Milne in distinguishing three principal phases of earthquake-motion—the preliminary tremors, the principal portion and the end portion.

In the first phase, the period of the vibrations is usually one-fifth of a second or less; at Miyako (1896–98), the mean period was ·08 second; in one of the Mino-Owari after-shocks, the period was as low as ·023 second. In the principal portion, it ranges from half a second to a second, but occasionally reaches two seconds or more. In the earthquakes recorded at the Hitotsubashi (Tokyo) observatory (1887–89), the average period of the horizontal component was ·76 second and that of the vertical component ·53 second. In those recorded at the Hongo (Tokyo) observatory during the same years, the corresponding figures were ·57 and ·20 second. Slow undulations occurred in all three phases. At Miyako, their average period was nearly the same in each phase, namely, 1·0–1·1 seconds in the first, 1·0–1·3 seconds in the second, and ·94–1·3 seconds in the third. The average period of the ripples was also nearly constant, ·08 second in the first and third phases and ·10 second in the principal portion.

The range (or double amplitude) of the vibrations was usually less than 1 mm. In 366 earthquakes recorded at Tokyo during the years 1885–97, the range of the largest vibration was less than 6 mm. in all but seven shocks, and did not exceed half a millimetre in about 60 per cent. of the whole number. In five earthquakes, the maximum range was considerable, namely, 22·0, 28·4, 41·0, 73 and 76·0 mm. At Nagoya, the maximum range during the Mino-Owari earthquake of 1891 must have been about 223 mm. or 9 inches.

In slight earthquakes, the maximum acceleration is usually not more than 5 or 10 mm. per sec. per sec. The average for 64 earthquakes recorded at Hitotsubashi was 20 mm. per sec. per sec. In the Tokyo earthquake of 1894, the maximum acceleration was 444 mm. per sec. per sec. at Hongo and 900 at Hitotsubashi. From the dimensions of fallen columns, Omori estimated the maximum acceleration to be more than 4300 mm. per sec. per sec. at two places in the central area of the Mino-Owari earthquake of 1891*.

226. Some interesting measurements are those which relate to shocks that are just sensible on the one hand and to destructive earthquakes on the other. Omori concluded from many observations that an earthquake would be sensible if its range were to exceed ·007 mm. or its maximum acceleration 17·0 mm. per sec. per sec. In ordinary destructive earthquakes, he found the duration of the strongest part of the motion to vary from 4·0 to 13·9 seconds, with an average of 8·0 seconds. In extensive and violent earthquakes, such as the Mino-Owari earthquake of 1891, it might amount to as much as 28 seconds. In four local destructive earthquakes from 1894 to 1922, the motion, as he remarks in one of his latest papers, was quite simple, a well-defined preliminary tremor being abruptly succeeded by the maximum vibration, and this, again, by smaller movements. Though the epicentre was in no case more than 83 km. from the observing station, the maximum vertical motion was always small compared with the maximum horizontal motion (from $\frac{1}{12}$ to $\frac{1}{5}$). The earthquake of 1894 has been referred to above (art. 220), but it is remarkable that, in the three earthquakes of 1921–22, the direction of the maximum horizontal motion was approximately normal to the line joining the station to the epicentre, showing that the greatest movement was due to distortional vibrations†.

227. Under this heading, a brief reference may be made to Omori's measurements of the vibrations of various buildings. The papers in which they are described would fill a volume of

* For similar measurements made in this and other earthquakes, Omori drew up his absolute scale of intensity for destructive earthquakes. It consists of seven degrees, the upper limit of the maximum acceleration for each degree being 300, 900, 1200, 2000, 2500, 4000, and more than 4000 mm. per sec. per sec. (*Publ.* no. 4, 1900, pp. 14, 137–141.)

† Tokyo, *Coll. Sci. Jl.* vol. 11, 1899, pp. 161–195; *Publ.* no. 5, 1901, pp. 1–82; no. 6, 1901, pp. 1–181; no. 10, 1902, pp. 1–102; no. 22 A, 1908, pp. 1–39; *Bull.* vol. 2, 1908, pp. 206–209; *Notes*, no. 5, 1923, pp. 1–7.

some 700 pages*. Omori made many experiments on the fracturing and overturning of brick columns fixed to a shaking table (1893–1910), he measured the vibrations of brick buildings during earthquakes (1900–08), also those of railway bridges and their piers (1902–10), of railway carriages (1904–10) and torpedo-boats (1904), and, lastly, of various towers and chimneys, two of the latter being of reinforced concrete and rising to heights of 567 and 660 feet (1902–21). In most cases, the object of the measurements was to ascertain the best forms to be given to the structures to enable them to resist the shocks of earthquakes. To this extent, they lie outside the range of this volume, but it may be of interest to notice one point. In the lateral vibrations of railway carriages, the maximum acceleration was sometimes as much as 2000 mm. per sec. per sec., or equal to that experienced during destructive earthquakes at San Francisco in 1906 and Messina in 1908.

228. Duration of Preliminary Tremor. At an early date, Omori sought for a formula connecting the epicentral distance (x km.) with the duration (y secs.) of the preliminary tremor. Plotting the points with distance and duration as coordinates, he found them lying roughly along a straight line. He therefore assumed that x and y are connected by a linear relation

$$x = ay + b,$$

and, by the method of least squares, determined the values of the constants a and b from a number of corresponding values of x and y†. In his first paper (1899), he found the relation to be

$$x = 7{\cdot}51y + 24{\cdot}9$$

for values of x between 100 and 900 km. He also showed how, from the observed durations of the tremor at several stations, the position of the epicentre could be determined. The formula

* *Publ.* nos. 4, 9, 12, 15 and 20; *Bull.* vols. 1, 2, 4, 9, etc. Such measurements had been suggested by J. D. Forbes in 1841 (art. 44) and were carried out by Milne in railway carriages in 1889 (*Japan Seis. Soc. Trans.* vol. 15, 1890, pp. 23–29).

† In his second paper on this subject (*Publ.* no. 5, 1901, pp. 58–70), Omori assumed that the linear relation $x=7{\cdot}43y+24{\cdot}9$ held for all distances from 100 to 10,000 km., and this led him to conclude that the waves of the first preliminary tremor are transmitted nearly parallel to the surface of the earth and at a certain (probably constant) depth below it.

was modified in 1901 and 1903. For local earthquakes, he found

$$x = 7{\cdot}27y + 38$$

for values of x between 70 and 900 km., changed in 1908 to

$$x = 6{\cdot}86y + 8{\cdot}1$$

for values between 50 and about 200 km. A still simpler relation

$$x = 7{\cdot}48y$$

was used for neighbouring volcanic earthquakes, and this was finally replaced by

$$x = 7{\cdot}42y$$

for values of x less than 1000 km.*

229. As Omori points out, this formula gives the focal, not the epicentral, distance; and thus, if the latter is otherwise known, the depth of the focus can be calculated. In this way, using the formula $x=7{\cdot}48y$, he found the mean depth of 729 non-eruptive earthquakes of the Asama-yama in 1911–12 to be 4·8 km. below the top of the mountain or about 3 km. below its base†. In the last year of his life, using the formula $x=7{\cdot}42y$, Omori determined the focal depths of 21 earthquakes felt in Tokyo during the years 1919–22. The earthquakes were of various intensities, strong, moderate or slight, and all originated in the provinces adjoining Tokyo and Sagami Bays. The depths varied from 16 to 49 km., the average value being 33 km.‡

230. Japanese Earthquakes. The great catalogue of 1898 earthquakes compiled under Sekiya's direction has been referred to above (art. 220). It was issued with notes by Omori in 1899. He estimates that, down to the end of the preceding year, there were 222 destructive earthquakes in the whole of Japan, and 108 since the beginning of the seventeenth century. Thus, a great earthquake occurs on an average once in every 2¾ years. A second catalogue of the stronger earthquakes from 1902 to 1907

* Tokyo, *Coll. Sci. Jl.* vol. 11, 1899, pp. 147–159; *Publ.* no. 5, 1901, pp. 58–70; no. 13, 1903, pp. 86–96; *Bull.* vol. 2, 1908, pp. 144–147; vol. 9, 1918, pp. 33–40.

† *Bull.* vol. 6, 1912, pp. 127–128, 238–239.

‡ *Bull.* vol. 11, 1923, pp. 24–27; *Notes*, no. 2, 1922, p. 11; no. 3, 1922, pp. 12–13; no. 4, 1924, p. 9.

was compiled by Omori in 1908. In these six years, the total number of earthquakes recorded in Japan was 9628, of which 621 disturbed land-areas ranging from 600 to 120,000 square miles*.

The first earthquake that Omori studied in the field was the Mino-Owari earthquake of 1891. In this, he confined his attention to two points, the intensity and the direction of the movement. From a great number of overturned stone-lanterns and other bodies, he determined the maximum acceleration at many places by means of the formula $f=xg/y$, where y is the height of the centre of gravity above the base and x its horizontal distance from the edge about which the body was overturned. The highest values of the acceleration exceeded 4300 mm. per sec. per sec. These measurements had two important results. It was on them that Omori based his absolute scale (art. 225), and by their means that, for the first time, isoseismal lines corresponding to absolute intensities were drawn on an earthquake-map. In his map of the central area, the zone of extreme violence is shaded, nearly the whole of the isoseismal of 2000 mm. per sec. per sec. is traced, while portions of that of 800 mm. per sec. per sec. are indicated. The fall of well-shaped stone-lanterns provided numerous measurements of the direction, the average directions at 43 places being shown on the map. Omori found that these directions were approximately normal to, and directed towards, the meizoseismal zone†.

Of the Tokyo earthquakes studied by Omori, the strongest was that of 20 June 1894. The remarkable seismogram of this earthquake was described by Sekiya and himself (art. 220), and Omori's subsequent papers relate to the directions of fall of 245 stone-lanterns and other columns. The mean of all the directions (S. 71° W.) is practically identical with the direction of the maximum horizontal motion. In three of his latest notes, Omori describes the severe or strong earthquakes of 8 Dec. 1921, 26 Apr. 1922 and 14 Jan. 1923, and, from the durations of the preliminary tremors and the known epicentral distances, found

* Tokyo, *Coll. Sci. Jl.* vol. 11, 1899, pp. 389–437; *Bull.* vol. 2, 1908, pp. 58–88.

† *Ital. Soc. Sism. Boll.* vol. 2, 1896, pp. 189–200; *Publ.* no. 4, 1900, pp. 13–24.

the depths of the foci to be, respectively, about 29, 48 and 48 km.*

231. Distribution of Japanese Earthquakes. Milne in 1884 had shown how large a proportion of Japanese earthquakes were of submarine origin, and in 1895 he pointed out how frequently destructive earthquakes occurred under the steep oceanic slopes on the eastern side of the islands (art. 201). With ampler materials, Omori insisted on the same law of distribution. Dividing the 222 destructive earthquakes since the fifth century into two classes, local and non-local, according as they affected one province or several provinces, he showed that there were hardly any but local shocks on the Japan Sea side, while the Pacific side was often disturbed by great non-local shocks with origins situated beneath the ocean. Of the ten most extensive and violent earthquakes, three originated in Central Japan and seven off the south-east coast. On the Japan Sea side, five earthquakes were followed by small sea-waves; on the other, great sea-waves swept in after 23 earthquakes.

Again, during the years 1885–1905, 257 earthquakes in Japan disturbed areas of more than 25,000 square miles. Of these, 145 had their origins in the great "external seismic zone" off the east coast of the islands, and nine in the "inner seismic zone" off the west side. Four other zones were prominent—the valley of the Shinano river (16 earthquakes), Tokyo Bay and Sagami-Nada (the seat of the great earthquake of 1 Sep. 1923, 12 earthquakes), the provinces of Mino, Owari and Echizen (12 earthquakes), and the western half of the inland sea (23 earthquakes†).

Of the special districts studied by Omori, the most interesting are the valley of the Shinano and the country round Tokyo. The Shinano valley runs for some distance parallel to the west coast, close to it, and lying about north-west of Tokyo. During the last century, it was visited by three destructive earthquakes, the first around Sanjo on 18 Dec. 1828; the second, on 7 Dec. 1833, in the

* *Ital. Soc. Sism. Boll.* vol. 2, 1896, pp. 180–188; *Publ.* no. 4, 1900, pp. 25–33; no. 22 A, 1908, pp. 1–39; *Bull.* vol. 1, 1907, pp. 194–199; vol. 2, 1908, pp. 7–12; *Notes*, no. 2, 1922, pp. 1–21; no. 3, 1923, pp. 1–30; no. 4, 1924, pp. 1–9.

† Tokyo, *Coll. Sci. Jl.* vol. 11, 1899, pp. 413–418; *Bull.* vol. 1, 1907, pp. 114–123.

Sado island and the adjoining coast districts, with its epicentre about 115 km. N.N.E. of the former; and the third, one of the most violent of all Japanese earthquakes, on 8 May 1847, with its epicentre close to Nagano and about the same distance south-west of the Sanjo epicentre. Besides these destructive earthquakes, there have recently (1886–99) been five strong earthquakes in the same zone. In all of them, the epicentres lie outside the meizoseismal areas of the three destructive earthquakes; and this, in Omori's view, is in accordance with the principle that great earthquakes are not repeated in one and the same centre*.

During the eight years 1914–21, 199 earthquakes were felt in Tokyo. With a few exceptions, they originated in four well-defined zones—the eastern half of the Boso peninsula and the adjoining sea-bed, the northern half of Sagami Bay and the country beyond, the district round Mount Tsukuba to the north and north-east of Tokyo, and a submarine band lying off the east coast of the Main Island. Between the first three regions lies a district occupied by a portion of Sagami Bay, Tokyo Bay, and the low Musashi plain to the north. That it was almost immune from earthquakes during these years, Omori attributes to the relief afforded by the Tokyo earthquakes of 1855 and 1894. In the course of time, however, he remarks, the three zones "will become gradually quiet, while the Musashi plain and the Tokyo bay may, as a compensation, recommence its seismic activity, and may result in the production of a strong earthquake, probably just after a year of marked minimum seismic frequency." It is remarkable that the district here indicated should lie in the northern part of the epicentral area of the great earthquake of 1923†.

232. Periodicity of Japanese Earthquakes. No country in the world can vie with Japan in the possession of earthquake records of long duration. None, consequently, provides such ample materials for the study of seismic periodicity. Omori was chiefly interested in the annual and diurnal periods, though he attended to others of longer duration than either a day or a year.

* *Bull.* vol. 1, 1907, pp. 138–141; vol. 2, 1908, pp. 136–143.
† *Notes*, no. 2, 1922, pp. 14–16; no. 4, 1924, pp. 9–10.

It was to the annual period only that he gave his attention when discussing the Earthquake Investigation Committee catalogue in 1899. Taking the 222 destructive earthquakes from the fifth century onwards, he finds that the maximum frequency occurs in the summer months (June–Aug.). In ordinary small earthquakes which are not after-shocks of great earthquakes (1885–91), the maximum is in spring (Mar.–May) and the minimum in summer; that is, the annual variation of slight earthquakes is, roughly, opposite to that of strong earthquakes. He returned to this reversal some years later (1908), and showed that, both at Tokyo and Kyoto, the annual variation of small earthquakes is almost symmetrically opposite to that of destructive and semi-destructive shocks*.

The method adopted in the above discussions is the same as that used by Merian, Hoff and Perrey (more than half a century before (arts. 40, 41, 49). In his second paper (1902), the method applied was practically that of taking two-monthly means of the monthly numbers of earthquakes†. But—and this is the most interesting point— he considered the earthquakes, not as registered over the whole of Japan, but separately at 26 meteorological stations. At 15 of these stations, the maximum epoch of the annual period occurs in spring; and it is noteworthy that the earthquakes recorded were mostly of inland origin and that the annual variation of frequency follows that of barometric pressure. Of the remaining stations, eight gave the maximum epoch in summer. They lie chiefly in the north-eastern portion of Japan; the earthquakes are mostly of submarine origin and their annual variation of frequency is opposite to that of the barometric pressure on land‡.

The explanation of this curious distribution is given in two later papers on the annual variation in the height of the sea-level along the Japanese coasts. On both sides of Japan, the annual fluctuation in barometric pressure is opposite to that in the sea-

* Tokyo, *Coll. Sci. Jl.* vol. 11, 1899, pp. 403–408, 423–426; *Bull.* vol. 2, 1908, pp. 17–20.

† In 1884, C. G. Knott (art. 190) had applied the more efficient method of overlapping means (six-monthly means) in the investigation of the annual periodicity of earthquakes.

‡ *Publ.* no. 8, 1902, pp. 1–94.

level. But—to take one station, Miyaki, as an example—the annual fluctuation in sea-level was 276 mm., while that of the atmospheric pressure was equivalent to 126 mm. of water. Thus, the total pressure on the sea-bottom is greatest in the summer months, and the frequency of submarine earthquakes follows, on the whole, the variation in the total pressure on the sea-bed*.

Omori's discussions on the diurnal periodicity of earthquakes and on various fluctuations in after-shocks frequency are less convincing than his treatment of the annual periodicity. He concludes, however, that variations in atmospheric pressure are probably responsible for the diurnal fluctuations in earthquake-frequency†.

233. Earthquakes of Formosa and other Countries. The island of Formosa became part of the Japanese empire in 1895, and, in the following year, the first seismograph was installed at Taihoku. Omori's most important memoirs on the Formosan earthquakes were published in Japanese only‡, but he wrote six notes or preliminary reports in English. The most interesting of these relates to the destructive earthquake of 17 Mar. 1906. The earthquake was an unusual one, for no other is known to us in which the direction of horizontal displacement varied. At the east end of the fault (so far as it could be traced), the south side was lowered 6 ft. and shifted 6 ft. to the west. Elsewhere, the north side was depressed 2–4 ft. and moved 2–8 ft. to the east. The continuation of the fault-line passes through the epicentre of the earthquake of 11 Jan. 1908, an earthquake which belongs, however, to the longitudinal seismic zone that runs near the east coast for more than half the length of the island. The recent earthquakes of this zone follow a law of distribution to which Omori more than once drew attention. The Giran earthquake of 1901 occurred near the north end of the zone, the Karenko earthquake of 1903 near the south end, the Taito earthquake of

* *Publ.* no. 18, 1904, pp. 23–26; *Bull.* vol. 2, 1908, pp. 35–50.

† Tokyo, *Coll. Sci. Jl.* vol. 7, 1894, pp. 126–138; *Publ.* no. 8, 1902, pp. 53–94. See also Tokyo *Phys. Math. Soc. Reps.* vol. 2, no. 8.

‡ *Jap. Publ.* no. 54, 1906, pp. 1–223; no. 88 B, 1919, pp. 62–71.

1905 between them but nearer the north end, and the earthquake of 1908 between the epicentres of 1903 and 1905*.

234. Outside the Japanese empire, Omori studied three great earthquakes—the Kangra earthquake of 1905, the Californian earthquake of 1906, and the Messina earthquake of 1908—in each case visiting the epicentral area. Of the Kangra earthquake, we have a detailed study of the seismograms†; on the Californian and Messina earthquakes only preliminary notes were published. The valuable monographs by the State Earthquake Investigation Commission on the Californian earthquake (art. 155) and by M. Baratta and others on the Messina earthquake perhaps made the complete reports unnecessary.

These preliminary notes are hardly representative of the long periods (about 2½ months) spent both in California and Italy. The most interesting point in the Californian note did not require a personal visit to the central area. It relates to the distribution of the earthquakes along the west coast of America. On 3 and 10 Sep. 1899 and 9 Oct. 1900, there were great earthquakes off the coast of Alaska; on 20 Jan. 1900 and 19 Apr. and 23 Sep. 1902, others in Mexico and Guatemala; on 31 Jan. 1906, one off the west coast of Panama, Columbia and Ecuador; and on 18 Apr. 1906, the Californian earthquake. As the epicentral areas of these four groups of earthquakes practically outlined the American coast from Alaska to Ecuador, Omori concluded that the next earthquake would take place at either end of this great zone, and probably at the south end in Chili and Peru. Early on 16 Aug. 1906, the prediction was verified by the occurrence of a great earthquake off the Aleutian Islands, followed, within half an hour, by another off Valparaiso‡.

The interest of the Messina note lies chiefly in Omori's estimate of the unnecessary loss of life. The destruction of the city, he says, "is really beyond one's imagination." Of its 150,000 in-

* *Bull.* vol. 1, 1907, pp. 53–69; vol. 2, 1908, pp. 156–165. For other papers on Formosan earthquakes, see Tokyo *Phys. Math. Soc. Proc.* vol. 2, no. 19, 1905, pp. 4–6; *Bull.* vol. 1, 1907, pp. 70–72; vol. 2, 1908, pp. 148–155; *Notes*, no. 5, 1923, pp. 9–16.

† *Publ.* no. 21, 1905, App. pp. 1–4 (3 plates); no. 23, 1907, pp. 1–16 (22 plates); no. 24, 1907, pp. 1–273 (11 plates).

‡ *Bull.* vol. 1, 1907, pp. 7–25, 26–43, 75–113.

habitants, he believed that one-half were killed. From the dimensions of overturned bodies, he found the maximum acceleration to be about 2000 mm. per sec. per sec. He contrasts the loss of life with that at Nagoya during the Mino-Owari earthquake of 1891. In this city of 165,000 inhabitants, only 190 persons were killed, though the maximum acceleration was 2600 mm. per sec. per sec. Omori thus concludes that "about 998 out of 1000 of the number killed in Messina must be regarded, when spoken of in comparison to a Japanese city, as having fallen victims to seismologically bad construction of the houses."*

235. Miscellaneous Phenomena. Under this heading may be included a few subjects on which Omori touched more or less briefly, such as the secondary causes of earthquakes, the attendant sound-phenomena, and the effects of the early tremors on pheasants, or which have little actual connexion with seismology like the movements called by him pulsatory oscillations.

As most Japanese earthquakes are, it seems, due to fault-formation, the final movement might be precipitated by local causes such as changes in the distribution of atmospheric pressure or in the height of the tides or in the amount of the rainfall. The jinari or earth-sounds of Arima, of which there were more than 1200 from 8 Aug. to 31 Dec. 1899, show four maxima throughout the day, and these coincide in epoch with the maxima and minima of atmospheric pressure. The non-destructive earthquakes recorded at Tokyo (2208 in number) from 1876 to 1899, also have four maxima in the day, the two principal maxima coinciding with the maxima, and the two secondary maxima with the minima, of atmospheric pressure. Of the earthquakes of moderate intensity (145 in number) felt in Tokyo from 1902 to 1907, those of inland origin were most frequent when the pressure was high, and those of submarine origin when it was low.

Again, of 14 strong submarine earthquakes (1890–1907), the mean difference in 13 between the time of occurrence and the nearest epoch of high or low water was less than an hour. Of the moderately strong earthquakes felt in Tokyo (1902–07), 65 were of inland, and 80 of submarine, origin. Of the former, 32

* *Bull.* vol. 3, 1909, pp. 37–45; vol. 5, 1913, pp. 99–100.

occurred with high water and 27 with low; of the latter, 32 with high, and the same number with low, water.

236. The common belief that earthquakes are frequent after storms of rain is, according to Omori, not without foundation. Though the curve of annual seismic frequency at Tokyo (1876–1907) seems to have no connexion with the amount of precipitation at the same place, it is almost identical in form with the curve of annual precipitation at Niigata and Akita, places on the north-western side of Japan and respectively 160 and 270 miles from Tokyo. The different maxima of frequency correspond and the curves are so nearly parallel that Omori concludes that the frequency of earthquakes in Tokyo varies with the amount of precipitation on the north-western side of the Main Island. He also thought that great shocks do not occur in bad weather or with violent storms, a fortunate exemption considering how often disastrous fires break out after great earthquakes in Japan*.

237. At various times Omori studied the sound-phenomena of Japanese earthquakes. Of the shocks felt on the loose soil of Tokyo, only 15 per cent. were accompanied by sound. On the other hand, on the hard rock of Mount Tsukuba, 93 per cent. of the sensible earthquakes were so attended; and, with those that originated within 20 or 25 miles, the sound was invariably heard. Thus, the audibility of the sound is greater on hard, than on soft, rock (art. 201); but it seems to be independent of the strength of the shock, for, of 35 earthquakes with loud sound felt on Mount Tsukuba, 10 were of moderate intensity, 21 were slight, and the rest unfelt†.

238. Pheasants have often been credited with a power of feeling tremors that are imperceptible to man. During the years 1913–16, Omori lived near a large garden in which were many pheasants. While working quietly in his study in the middle of the night, he noticed that, in half of the 22 cases of crowing, the pheasants were disturbed by movements that were insensible to a skilled observer‡.

* *Bull.* vol. 2, 1908, pp. 101–135.
† *Publ.* no. 22 A, 1908, pp. 1–39; *Bull.* vol. 11, 1923, pp. 33–63.
‡ *Bull.* vol. 11, 1923, pp. 1–5.

239. Microseisms or, as Omori called them, pulsatory oscillations are referred to in papers from 1899 to 1913. As is well known, they occur in storms that may last for several days. Omori notes the existence of three chief periods, the mean values of which are 2·9, 4·4 and 8·0 secs., and he compares the last two with the predominating periods in the preliminary tremors of distant earthquakes, the mean values of which are 4·6 and 8·3 secs. The oscillations with a mean period of 4·4 secs. are the more frequent, occurring in about 90 per cent. of the records. Such oscillations invariably herald the approach of deep barometric depressions. For those of longest period, the necessary condition seems to be a high pressure with a low gradient. That pulsatory oscillations differ from earthquake-vibrations is clear for several reasons: (i) though the mean periods at Hongo and Hitotsubashi (Tokyo) are the same, it is impossible to identify individual vibrations or maximum movements; (ii) the maximum amplitudes in both E.–W. and N.–S. components are equal; (iii) the direction of the horizontal oscillations is constantly changing; and (iv) the vertical movements are frequently in excess of the others. Pulsatory oscillations thus seem to be the result of underground disturbances at many points due to volcanic activity, the passage of deep barometric depressions, or the existence of heavy ocean swells*.

240. Volcanic Eruptions. The volcanoes, as well as the earthquakes, of Japan come within the scope of the Earthquake Investigation Committee. In one of his earliest papers Omori described the eruption of the Azuma-san in 1893, and, from that time, his interest in Japanese volcanoes never lapsed. In a paper published in 1907, he referred to the eruptions of the Unsen-dake in 1792, and, in one of the following year, he gave a list of recent volcanic eruptions in Japan†. But, during the last twelve years of his life, the interest of the volcanic phenomena seems to have prevailed over that of earthquakes, and the greater part of his time

* Tokyo, *Coll. Sci. Jl.* vol. 11, 1899, pp. 130–135; *Publ.* no. 5, 1901, pp. 51–57; no. 13, 1903, pp. 81–86; *Bull.* vol. 2, 1908, pp. 1–6; vol. 3, 1909, pp. 1–35; vol. 5, 1913, pp. 109–137.

† *Seism. Jl. Japan*, vol. 3, 1894, pp. 1–22; *Bull.* vol. 1, 1907, pp. 142–144; vol. 2, 1908, pp. 21–34.

was spent in the investigation of the eruptions of the Usu-san in 1910, the Asama-yama in 1908–14, and the Sakura-jima in 1914*.

The reports on these eruptions, and especially on those of the Asama-yama and Sakura-jima, are among our most important monographs on volcanic phenomena from a physical point of view. With many of the sections—those, for instance, on the course of the eruptions and the resulting changes of level, the mapping of the explosion sound-areas and the multiplication of the explosive reports—we have little concern in these pages. But, on the nature of volcanic earthquakes, on the relations between the eruptions and the accompanying tectonic and volcanic earthquakes, Omori's researches have added much to our knowledge.

(i) That tectonic earthquakes occur in close connexion with volcanic eruptions had been noticed by G. P. Scrope (1797–1876) in 1825†. Of this relation, Omori gives some interesting examples. About thirty hours before the eruption of the Usu-san on 25 July 1910, an earthquake occurred in that district that was felt to a distance of 87 miles from the volcano. In the south of Japan are three important volcanoes lying along a N.N.E.–S.S.W. line. The Kirishima-yama, at the north end of the line, broke out in strong eruption on 18 Nov. 1913; the Sakura-jima, 20 miles to the south, on 12 Jan. 1914; and the Iwo-jima, 45 miles farther southward, on 13 Feb. 1914. The great eruption of the Sakura-jima was followed in a few hours by an earthquake that damaged houses in Kagoshima and was recorded in European observatories. One month later, another strong earthquake occurred during the eruption of the Iwo-jima. There can be little doubt that, as Omori suggests, both earthquakes and eruptions were effects of the same deeply-seated subterranean action‡.

(ii) It has long been known that volcanic eruptions are preceded by numerous earthquakes, which become less frequent as the eruption begins§. Omori's observations on the Usu-san and Sakura-jima eruptions give precision to this statement. That of the former volcano began at 10 p.m. on 25 July 1910. At Nishi-

* Usu-san, *Bull.* vol. 5, 1911, pp. 1–38, 101–107; vol. 9, 1920, pp. 41–76 (21 plates). Asama-yama, *Bull.* vol. 6, 1912, pp. 1–257; vol. 7, 1914, pp. 1–456 (101 plates). Sakura-jima, *Bull.* vol. 8, 1914, pp. 1–525 (114 plates).

† *Considerations on Volcanos*, p. 155.

‡ *Bull.* vol. 5, 1911, pp. 15–16; vol. 8, 1914–22, pp. 22–25, 467–525.

§ Scrope, *Considerations on Volcanos*, 1825, p. 155.

Monbets, about 5 miles from the centre of the volcano, one earthquake was felt on 21 July, 25 earthquakes on 22 July, 110 on 23 July, 351 on 24 July, and 165 on 25 July. At Sapporo, about 44 miles from the volcano, the earthquakes were registered by a horizontal pendulum seismograph, there being one on 21 July, three on 22 July, 23 on 23 July, 76 on 24 July and 84 on 25 July. After the eruption began, the decline in frequency was rapid, the numbers being 26 on 26 July, and 15, 5, 6 and 1 on the succeeding days.

The eruption of the Sakura-jima began at 10 a.m. on 12 Jan. 1914. At Kagoshima, about 6 miles from the centre of the volcano, the first earthquake was recorded early on 11 Jan. After this, the hourly frequency of earthquakes was 4·1 from 3 to 11 a.m. on 11 Jan., 12·4 from 11 a.m. to 8 p.m., and 19·5 from 8 p.m. on 11 Jan. to 10 a.m. on 12 Jan. As soon as the eruption began, there was a marked decline in frequency, the numbers recorded during successive hours from 10 a.m. being 17, 11, 6, 3, 5, 2, 2 and 2*.

(iii) Omori's seismographic observations on the Asama-yama, in central Japan, show that there are two distinct types of volcanic earthquakes, one class being independent of any outburst of the volcano, while the others were invariably the results of explosions. They differed in several respects. The shocks without explosions consisted only of minute quick vibrations, those with explosions of slow movements, of as much as 2·6 and 5·3 seconds period, on which after a few seconds, quick vibrations were superposed. The earthquakes without explosions were stronger, but of shorter duration, than the others. Moreover, the two types of earthquakes alternated in frequency, the maxima and minima (both diurnal and annual) of the first type coinciding approximately with the minima and maxima of the second type†.

241. Conclusion. In the preceding pages, I have tried to give a summary of Omori's more important investigations. Long as it is, there are points on which I have not touched. Such are the elaborate measurements of the periods and velocities of the

* *Bull.* vol. 5, 1911, pp. 8–15; vol. 8, 1914, pp. 9–12.

† *Bull.* vol. 6, 1912–14, pp. 118–136, 227–257; vol. 7, 1914–19, pp. 91–165, 346–357.

different phases of the earthquake-motion, the tilting of the ground due to passing cyclones, the important investigations on the construction of earthquake-proof buildings. No other seismologist can equal, and few can approach, the enormous output of Omori's original work. Even if many of the measurements were made by others—and there can be no doubt that he owed much to his assistants—if the raw material were to some extent worked up before it passed under his hands, yet the careful study of such masses of observations must have involved incessant thought and labour. Often, as we learn from his notes on pheasant-crowing, his work was continued into the early hours of the morning, as if, like Keats, he feared that he might cease to be before his pen had glean'd his teeming brain.

Omori's work was naturally not without defects. If anything, it suffered from too great wealth of material. Memoirs were promised that were never written*, and, in at least two cases, preliminary notes supplied the place of the complete reports†. With so many subjects crowding in upon him and pressing for treatment, the time was too brief that was left for reflection, for the study of the previous literature. A more thorough mathematical training would have helped Omori greatly in much of his work, especially in his inquiries on seismic periodicity and in realising the limitations of his empirical formulae on the decline in frequency of after-shocks and on the relation between the duration of the preliminary tremor and the distance of the origin.

But such matters are as nothing when we think of his actual performance—of his studies of Japanese earthquakes and their distribution in space; of his investigation of various seismic periods and their causes; of his memoirs on the frequency, and the decline in frequency, of after-shocks; above all, of his great monographs on the eruptions of the Asama-yama and Sakura-jima. By his long-continued work on the construction of earthquake-proof buildings, he helped to deprive earthquakes of much of their destructive power; and, by his timely prudence, he saved many thousands of lives from the devastating eruptions of the Sakura-jima and Usu-san.

* *Publ.* no. 24, 1907, p. 273; *Bull.* vol. 1, 1901, pp. 44, 113.

† *Bull.* vol. 1, 1907, pp. 7–25; vol. 3, 1909, pp. 37–45.

INDEX

Printed in the United States
By Bookmasters